A
Slippery
Slope

A SLIPPERY SLOPE

The Long Road to the Breakup of AT&T

Fred W. Henck
and
Bernard Strassburg

CONTRIBUTIONS IN ECONOMICS AND ECONOMIC HISTORY,
NUMBER 80

Greenwood Press
NEW YORK • WESTPORT, CONNECTICUT • LONDON

Library of Congress Cataloging-in-Publication Data

Henck, Fred W., 1921-
  A slippery slope.

  (Contributions in economics and economic history,
ISSN 0084-9235 ; no. 80)
  Bibliography: p.
  Includes index.
  1. American Telephone and Telegraph Company--
Reorganization. 2. Telephone--United States.
3. Telecommunication--United States. 4. Corporate
divestiture--United States. 5. Antitrust law--United
States. I. Strassburg, Bernard, 1918-
II. Title. III. Series.
HE8846.A55H46 1988   384.6'065'73   87-28043
  ISBN 0-313-26025-7 (lib. bdg. : alk. paper)

British Library Cataloguing in Publication Data is available.

Library of Congress Catalog Card Number: 87-28043
ISBN: 0-313-26025-7
ISSN: 0084-9235

First published in 1988

Greenwood Press, Inc.
88 Post Road West, Westport, Connecticut 06881

Printed in the United States of America

The paper used in this book complies with the
Permanent Paper Standard issued by the National
Information Standards Organization (Z39.48-1984).

10 9 8 7 6 5 4 3 2

To Bettye Henck and Anita Strassburg

They have lived all this twice;
once when it happened and again in the retelling.

CONTENTS

PREFACE

To millions of Americans, one of the most perplexing questions of most recent years has been, "What in the world ever happened to our telephone system?"

The loss of a familiar part of our lifestyle, much like full-service gasoline stations, is something that has to be accepted, if not completely understood. Americans now diagnose their own telephone service problems and arrange to have them repaired or the offending parts replaced, much as most of them pump their own gasoline and check their own oil and tires. Without much to go on except for television advertising, they select their own long distance carrier or get "allocated."

Where once they could rely on the convenience of their "one-stop" local telephone company, now they are obliged to make choices formerly reserved for bigger-ticket items in family budgets.

A large part of the public found the drastic changes in the way they obtain telephone service to be sudden and bewildering. More than a few have echoed the words of the general manager of a National Basketball Association team. His club was widely reported, on the eve of the annual draft of college players, to be negotiating a trade to improve its position in the selection process. At the last minute, the deal fell apart. Asked why, he said, "It all goes back to the breakup of the phone company. Nothing has worked right since then."

These events, in truth, have been bewildering, but they were not sudden. The popular misconception—that most of the inconveniences and dislocation burst on the scene simultaneously with the January 1, 1984, breakup of the Bell System—does not stand up under examination. Divestiture was not an isolated event which by itself brought about the confusion and occasional chaos besetting the average consumer. Instead, it was the culmination of a process of change in telecommunications policy, which began several decades ago.

Advancing and new technologies, economic pressures, and social and political developments were the driving stimuli. But there was not central orchestration and no grand design by the nation's policymakers. If the change was not carefully planned, was it simply fortuitous? No, but it was a process of gradual, albeit spasmodic, evolution, rather than programmed revolution.

It is startling to set side by side the picture of the Bell System as it was when government attention first critically focused on it, in the 1930s, and a portrait of the way telephone service is organized today.

The Bell System of the past was a giant unparalleled in our national economic history. Through its nationwide holding company, the American Telephone & Telegraph Company, it owned or controlled the principal telephone company in each of the forty-eight contiguous states, providing local and long distance services in their respective territories. It handled almost all interstate long distance calls through the AT&T Long Lines Department. Its manufacturing company, Western Electric, made most of the country's telephone equipment and acted as the purchasing agent for items it did not itself produce. Its parts included the leading U.S. industrial laboratory, the Bell Telephone Laboratories. In short, it was the dominant technological and operating force in the nation's telecommunications infrastructure.

It was regarded, domestically and abroad, as operating a model telephone system. By most standards, AT&T was the world's largest private corporation. It had the most employees. It was the largest consumer of privately held capital. It was often at the top of the list in gross revenues and net income. The divestiture of 70 percent of its assets as of January 1, 1984, ranks as the most colossal "instant" breakup of a corporate institution in American history.

Our story opens with the creation of the Federal Communications Commission (FCC) in 1934. There had been interventions by the federal government into the structure and policies of the Bell System before then, but they were infrequent, and the Bell management had accommodated them with minimum impact on its style and growth.

The sometimes open warfare between competing telephone exchanges, which erupted after the basic Bell patents expired, had led to a policy of self-restraint and careful balancing, but only after the Bell System dominated the cores of the nation's major population centers. By agreement with the Justice Department, the Bell companies acquired "independent" territory only when others were not ready or able to serve it. The telegraph service of Western Union which had been acquired earlier was divested, but the Bell System retained for many years a form of telegraph service, teletypewriter exchange or TWX, which operated more like the telephone system.

With great success and for many years, the Bell system fostered the concept that technologically and economically, local and long distance telephone services had the characteristics of a natural monopoly. This in turn meant the exclusion of directly competing suppliers of service or equipment, as well as undivided responsibility for end-to-end service as the only way it could properly be provided.

Telephone companies, therefore, did not lease instruments or other equipment. They provided a complete service directly to users; middlemen, such as brokers or resellers, were not tolerated.

Government regulators at all levels embraced the natural monopoly concept as the essential framework of their policies. State utility commissions— and at the federal level, first, the Interstate Commerce Commission, and for many years to follow after its establishment, the FCC—merely monitored the level of rates and earnings. They did not interfere with the prerogative of management to make basic business and operating decisions.

It was truly a symbiotic relationship. The regulated monopoly operated in what was considered to be the public interest and, in turn, was shielded against incursions by rivals and competitors, including the possibility of government ownership.

Very slowly, often haltingly, the system began to change. Public and political criticism of the Bell System's increasing dominance and size, on the one hand, and regulatory passivity, on the other, slowly built into challenges which, in turn, provoked changes in policy. Gradually these policy changes acquired momentum and eventually accelerated pell-mell into the present structure we all know. What happened to the comfortable and widely accepted institution, *the* telephone company, and why did it happen as it did? That is the focus of this book.

For the answer, one must look to a variety of sources. First was a federal agency, the FCC, which was created by Congress in 1934 as an administrative move to consolidate all governmental responsibilities regarding electrical communications within a single agency. Its primary mission, however, seemed to the public until the middle of the twentieth century to consist almost entirely of the licensing and regulation of broadcasting stations. Because of its statutory mission to oversee communications matters, the FCC was almost always the focal or departure point of the policy changes that have come about.

Responding to an exploding economy and technology and to the demands of the populace seeking to communicate computerized data as well as voice, the FCC assumed an increasingly activist role in its relationship to the Bell System. Slowly, and then with the pace gradually picking up, it turned from pro-monopoly policy to competition, as a more efficient way to deliver products and services in selective communications markets.

One must also take into account the amorphous role played by Congress —the ostensible fountainhead of national communications policy. It never passed its promised "basement to attic" rewrite of the Communications Act, but it came so close for so long, and its efforts were so convincing, that it served as a galvanizing agent in compromise policy changes administratively fashioned by the FCC and acceptable to diverse forces which could not agree on the precise terms of legislation.

It is necessary to remember the sometimes crucial role taken by a number of federal judges who, on occasion, did more to influence the direction of government policy than either the FCC or Congress. By affirming, modifying, or rejecting administrative decisions subject to furious assault, they made their own contributions to the framework that evolved.

It was a militant group of antitrust specialists in the Department of Justice who concluded that the time had come to restructure the communications industry more in keeping with a competitive environment. It was their initiative that mobilized the legal machinery to produce the ultimate result.

Finally, and most significantly, the responses of the industry itself were critical to the outcome. The Bell System often chose not to yield any part of its monopoly domain to the new entrants or to adjust its policies to the advent of a competitive marketplace. Eventually, in the face of the forces arrayed against it, this led to its dissolution.

Although the incredible metamorphosis of the telephone industry now seems, with the benefit of perfect hindsight, to have been close to inevitable, it was not that way as events transpired. Those policymakers at the FCC who believed that introducing modest increments of competition in selected markets would be beneficial never contemplated that it would lead to the industry's total transfiguration. The initial moves adopted by the agency were intended to serve as a curb or check on the growth of monopoly and to provide some marketplace standards as a helpful ally for more effective regulation.

Once started, change gathers its own momentum. The regulated Bell System was an excellent instrument to serve most of the public's communications needs and desires. But, like all human institutions, it was not perfect, and it could not forever be all things to all people.

Probably it was a creature of its times—times on which we look back fondly, but cannot return to. Big as it was, it could not contain within its borders all ideas, all innovations, all technology, all investment capital, and all eager entrepreneurs and thinkers. The society it served was too large. There had to be room for the "outsiders"—a reality it would not acknowledge until it was too late.

The early suggestions for change, the initial questions, the gradual infusion of a small amount of competition, and AT&T's resistance to change, all moved inexorably through a clamorous public debate to the crashing climax of divestiture. The merger of a few cells finally produced an all-encompassing giant.

Having been allowed the opportunity to be on hand from the early years of the story and having taken part in many critical stages, each of us has often been asked why we did not write a book about the telecommunications policy evolution. As contemporaries, longtime friends, and close associates during the many years our workweeks paralleled, we felt this collaboration was the proper response.

For reasons which should become clear as the story unfolds, we will occasionally revert to individual authorship, duly labeled. There were experiences and there are perspectives and viewpoints which cannot be shared.

Most of the time, we have been able to meld out divergent outlooks into a mutually acceptable joint authorship. But a government official and a reporter cannot be expected to agree on everything!

We are indebted to Beverly Rector and Norma Dawson for their patience and skill in typing this manuscript and to the law firm of Ward and Mendelsohn for so generously providing us with the support of their services and facilities. We are also grateful to the publishers of *Telecommunications Reports* who permitted our unrestricted access to their file of back issues, which was invaluable as the only available reference source for much of the story we have to tell. Above all, we are gratified that our friendship endured the rigors of this joint undertaking.

A
Slippery
Slope

1

THE WAY IT WAS

A bored group of congressmen ringed the ornate committee room rostrum as the incumbent chairman of the Federal Communications Commission insisted on devoting a section of his prepared statement to the FCC's common carrier (non-broadcast) activities, and doggedly read every word of it.

The House Interstate and Foreign Commerce Committee, legislative overseer of the FCC, was conducting its annual "oversight" hearing. The congressmen, temporarily putting aside the threat of World War II, were turning to more mundane matters. But not all that mundane to House members, who used the radio to keep in touch with their constituents and run for reelection, was what they considered the centerpeice of the Commission's work—its regulation of broadcasting stations.

After all, the telephone was just a business tool, whose reliability and constant availability they took for granted. So why spend a lot of time on common carrier issues before the FCC?

As the FCC chairman finished his statement, a senior member of the committee stirred restlessly and prepared to resume the questioning. "All right," he said, "so much for that. Now let's get back to the important subject—radio stations."

Some years later, the committee's membership had changed considerably. At the same time, there were faint stirrings of congressional interest in the common carrier area, and now and then constituents raised questions about it. This time, the full committee's Communications Subcommittee had called the FCC to Capitol Hill to talk about the issues of the day.

As usual, the FCC chairman had prepared a fairly lengthy statement on complex common carrier subjects, which he proceeded to summarize for the legislators. The newly appointed chairman of the House subcommittee leaned forward in his chair, evidently listening intently.

After taking in a long explanation of a current case involving the Bell System, the subcommittee chairman spoke for the first time. "There is," he said in a puzzled tone, "one thing about your statement I don't understand." Eager to make his points to an influential congressman, the FCC chairman replied briskly, "Yes sir. I'll try to clear it up."

"Tell me," the chairman of the House Communications Subcommittee asked in an official voice, " is AT&T part of the Bell System?"

That was one question about AT&T, parent, supervisor, and owner of most or all of the stock of the Bell telephone companies around the country, that all the members of the FCC could answer. Whether some of them could have answered very many more without consultation with their staff was an open question.

The huge Democratic congressional majority that descended on Washington after Franklin Delano Roosevelt's landslide victory in 1932 agreed on one thing to do as their part in launching a New Deal for a depression-battered nation. They agreed that most businesses essential to the economy needed a lot more government supervision.

The eager New Deal architects who swept into Washington with them, all reformers at heart, took especially long, hard looks at industries with monopoly and public service characteristics. In relatively short order, they had pushed through Congress a series of laws providing for tighter business controls and establishing or expanding and revamping regulatory commissions to exercise that supervision, including the federal Securities and Exchange, Power, Trade, and Communications commissions.

The FCC, which formally came into being July 11, 1934, was a packaging of the powers of two former regulatory commissions tied together with wonderfully resilient elastic. The Interstate Commerce Commission (ICC) had been "regulating" communications common carriers since before World War I, and the common carrier sections of the Communications Act were taken substantially from the governing law of the ICC, whose principal business was overseeing trucking, railroad, and bus companies.

The Federal Radio Commission (FRC), which went out of business with the launching of the FCC, had been set up in 1927. In a half dozen years after the start of commercial broadcasting by KDKA in Pittsburgh in 1921, the broadcasting industry had gone totally out of control. Stations went on the air willy-nilly at the whim of entrepreneurs, picked whatever frequency they wished to use, and broadcast in a way to interfere with the signals of others using the same slot in the radio spectrum. It was mainly to serve as "policeman" to halt the chaos in the airwaves that the FRC was set up, largely under the guidance of then-Secretary of Commerce Herbert Hoover.

The Communications Act of 1934 gave the new FCC a wide-ranging set of powers over the fledgling radio business. This was the job of the agency that drew most attention, caused most of its problems, and resulted in frequent trips to the presidential and congressional woodshed. The power to choose between applicants for profitable slots in the public's airwaves was a difficult one to administer comfortably.

The common carrier sections of the new Communications Act may have come in large measure from the old Interstate Commerce Act, but the governmental apparatus it set up was changed considerably. One significant difference was that now a special agency had been set up to regulate the communications companies, to replace the transportation-oriented ICC in that field. Another was that administrative powers of the FCC were more specifically delineated and spelled out.

Perhaps most important, there was a certain genius in the drafting of the law which did not become completely clear for many years. At the beginning of the act, the FCC was given broad and generally stated goals of advancing the "public interest" in available, efficient communications services at reasonable charges. Over the ensuing more than half century, the Communications Act has been—without substantial change, despite many congressional attempts—the vehicle for policy-making cycles ranging from fairly tight control of the industry to the modern "free market" deregulatory fetish. If it is careful to follow due process, the FCC can pretty much control or decontrol as it sees fit, and the courts for the most part have upheld this latitude.

Cynics may suggest that the all-purpose utility of the law came about because Congress really did not devote any unusual amount of attention to it. Committee hearings and floor debate which led to passage of the Communications Act were relatively brief and fairly routine. Lawyers searching in later years for legislative history and congressional intent have found the pickings fairly slim.

The subject of most congressional attention was assurance, vigorously sought by state public utilities commissions and their representatives, that the new FCC would not intrude on the fixing of rates for local and intrastate telephone service. Specific language to that effect was included in the act. Local ratesetting has remained a state province, although later litigants have learned in the federal courts that the wall around the states shutting off other federal policies was not nearly as impervious to federal incursions as their forebears thought it was in 1934.

The New Deal sought to end the days of rather casual govenment intervention in businesses controlled by "tycoons," and the FCC clearly was part of the mechanism. Its seven commissioners (much more recently, five) had various political connections, but relatively little experience for the job they were facing. One of the initial seven, former Oklahoma Corporation Commissioner Paul A. Walker, had experience in regulating telephone companies and was the "token" member with experience in the common carrier field.

Accounts of the FCC's early years make it plain that licensing and regulating broadcast stations not only was seen as the Commission's major job, and received what was estimated to be 90 percent of its time and attention, but caused most of its difficulties. The agency was under almost constant congressional attack from the beginning and was threatened with reorganization frequently, but the objections to what it was doing were focused almost exclusively on broadcasting.

The telephone industry was not entirely neglected. With congressional and administration reformers making studies, enacting legislation, setting up new agencies, and bringing large numbers of eager lawyers, economists, and other professionals to Washington, the telephone business was included with the other prominent national industries to come under surveillance.

Nonetheless, public concern about any monopoly depredations of the telephone companies was much less focused than the spotlight turned on "robber barons" in other utility industries. Telephone service was generally good, and there were no financial collapses in the telephone industry wiping out thousands of stockholders, or disclosures of stock manipulation, fraud, or "watering." The telephone business was still a monopoly, however, and therefore the launching of the new FCC was accompanied by a congressional resolution directing the Commission to make what became known as the Special Telephone Investigation (conveniently known as "the investigation") and to report its recommendations to Congress.

Emphasis in the resolution and a later FCC order setting up the investigation was on AT&T control of the local Bell operating companies—through control of financing; with the license contract, by which the local companies paid for centralized services rendered from AT&T headquarters; and by interlocking boards of directors. Emphasis was placed also on the standard supply contract, under which the AT&T-owned Western Electric Company was the Bell System's principal equipment manufacturer and its purchasing agent for material bought from other suppliers. The congressional resolution also referred to the Bell System's failure to reduce rates "generally" during the "years of declining prices." Congress had trouble referring to the Great Depression of the early 1930's by name.

"Big Business" was commonly blamed for the depression, and the eager students and young professionals who applied for the 100 jobs allocated to the investigation shared the zeal for reform. The Commission interviewed hundreds of applicants for the jobs. In those years, when just about any job at all was considered "good," government jobs were "very good." Some paid as much as $4,000 or $5,000 a year, and the ultimate was to become a commissioner at the incredible rate of $10,000 per year, or one of the three top FCC staff officials at $9,000.

The Commission paid the top staff salaries to its general counsel and chief engineer, but had considerable trouble getting the approval of an increasingly economy-minded Congress to compensate a chief accountant at the same figure. Recruiting personnel was nonexistent as a problem. One could live comfortably in Washington on even the lower government salaries at a time when a full-course dinner in a family restaurant was available for sixty-five cents, and a dime tip was the norm; haircuts were twenty-five cents or at most thirty-five cents; and transit fare was a dime, or less if the rider bought tokens. For those few who owned cars which sold new for as little as $500 or $600, parking spaces were easy to find on the streets.

Congress appropriated $1,525,000 for the FCC's first year and threw in a then-astronomical extra fund of $750,000 to finance the entire investigation. (In recent years, the FCC's annual appropriation has been in the $90 million to $95 million range). As with virtually all government operations, the $750,000 fund ran out prematurely, and the FCC, again with difficulty, finally got another $400,000 to finish the job. A few years later, it required several annual drives, the first two rejected, before the FCC could get a "deficiency" appropriation of $140,000 to wind up a few investigation-related projects.

The statistics of what was viewed then as a giant monopoly look small by comparison today. At the end of 1934, there were 13,458,000 telephones in service, compared with about 211,500,000 at the close of 1985. The total was up by only 38,000 Bell telephones two and a half years later, in mid-1937, but by then the Bell System had regained nearly half of the 17 percent decline from the 1929 high. The "telephones in service" figures included home extensions and all telephones on business premises. A large proportion of residence phones were on two-, four-, and eight-party lines. A special 1937 census, ostensibly to aid in business recovery but largely a bit of government "pump-priming" to increase employment, showed one telephone for every seven U.S. citizens. Today, because business and residence phones are combined in the statistics, the ratio is essentially one for one.

From the point of view of money, the Bell System's numbers were equally unimpressive by today's standards. For example, AT&T's (and the Bell System's) net income in 1935 was $145 million. Until the January 1, 1984, breakup of the Bell System, the three-month quarterly net income figures usually topped $1 billion. In 1935, AT&T paid $42 million in quarterly dividends to about 675,000 shareowners. In 1983, right before divestiture, the more than 3 million AT&T shareowners usually received $1.3 billion every calendar quarter.

On the consumer side, in the interstate area regulated by the FCC, long distance rates declined steadily in dollar charges, even though the cumulative totals of reductions reported were minuscule by today's yardsticks. Voluntary or negotiated long distance reductions were the order of the day, as they continued to be for three decades. As general business conditions slowly improved, although the nation never really emerged from the depression until the frenzied activity of World War II, AT&T found it feasible to negotiate rate cuts rather than go to confrontational formal hearings. The FCC staff may have been ready if not eager for a formal confrontation, but the commissioners knew the political value of being able to report that their efforts had led to still more rate reductions.

Soon after the FCC went into business, AT&T signaled that it understood the rules of the road by agreeing to a $3 million interstate rate reduction. At the end of July 1936, several months after the first hearings in the telephone investigation, a $7.35 million rate reduction was negotiated. For example, the basic station-to-station rate, similar to today's direct dialed call, was cut to $7.50 from $9 for a three-minute New York-to-San Francisco call. Near the end of the same year, another $12 million reduction, this time mostly on calls up to 234 miles, was announced. By March 1940, the New York-to-San Francisco rate was down to $6, and it dropped at that time to $4.50 as the result of another $5.5 million rate cut. The first night and Sunday rates had gone into effect by then, and the same call became $3, instead of $4.25, during those time periods.

How could dollar rates go down so far so fast? One can cite economic, political, and technological factors. The use of the long distance telephone was becoming more commonplace and routine, in business at least, and while calling volume went up, costs per call were starting to come down. Long distance was still a relatively small part of the Bell System's business, and the reductions had little effect on AT&T's ability to pay the $9 per share annual dividend rate, unbroken during the depression even when payments were often made in part out of accumulated surplus (during that time, AT&T stock usually cost $100 to $150 a share on the market).

Certainly no one in the telephone business was getting obscenely rich. It was pointed out at the end of the investigation that AT&T's earnings, from all sources, had averaged 6.75 percent on its net investment—the total capital in the business less the depreciation and amortization reserves. In 1936 all the carriers under FCC jurisdiction—essentially the larger telephone companies and the domestic and international telegraph companies—filed reports on all officials making $10,000 a year or more. The average salary for a board chairman was $45,833, and for a company president it was $33,000. The entire U.S. telecommunications industry had only three officials making $62,500 a year or more.

What was happening, foremost among other things, was AT&T's understanding that it made far more political sense to trim back its growing revenues modestly than it did to get into a formal rate case with the government. In World War I, AT&T officials recalled with never-failing concern, the Post Office Department took over nominal control of the telephone industry, although telephone people continued to run it. And as recently as 1928, bills were introduced in Congress that would legislate government ownership of the nation's telephone and telegraph industries.

During the New Deal, and perhaps for a little while thereafter, number one in all Bell System considerations was never to do anything that might lead to outright government ownership, something that could be plainly seen on the rise in a number of nations. There was never any such serious move in the United States, but the worry was there for Bell. AT&T officials never forgot a statement in the investigation staff, or interim, report which went to Congress April 1, 1938. It observed that, in the absence of regulation, government ownership would be almost the only means available to insure telephone service at low cost. The "Walker report" (named for Commissioner Walker, who presided over the investigation) added, "As a practical matter, government ownership would not be as difficult of attainment as the reestablishment of effective competition." Clearly, it was easier and better for AT&T to help make regulation work.

When the investigation was ordered, formal hearings were to start early in 1935. As it turned out, the FCC took a lot longer to fill the top staff spots and prepare for the sessions than had been expected, and the opening hearing did not take place until more than a year later, in mid-March 1936. Meanwhile, apparently still getting a handle on its common carrier responsibilities, the Commission held extensive hearings in two other areas which soon would be considered fairly routine.

First, it seemed to come as a surprise to some that the major communications companies, which usually owned 100 percent or at least a large majority of the stock of their operating subsidiaries, controlled them. The heads of AT&T, ITT, RCA, and Western Union and other top officials were called to Washington for lengthy testimony on the intracompany relationships. Then the FCC decided that those officials would not be allowed to sit on the boards of directors of the subsidiaries. There was no indication that this particularly inhibited AT&T President Walter S. Gifford, for example, from making the final decisions on Bell System-wide policy. That, it was generally believed, was what he was being paid to do.

Second, AT&T filed, in the FCC's early days, an application to build the first coaxial cable, from New York to Philadelphia. The FCC flexed a few regulatory muscles before giving its approval to the application. But the new high capacity cables, along with the later blooming microwave radio and finally satellites, were destined to become the backbone of the national telephone and television transmission networks.

At the outset of the special telephone investigation, its leadership made a decision which was to plague the FCC for years thereafter. It was decided to run the hearings like those of a congressional committee, since Congress had ordered the probe. AT&T lawyers were not allowed to conduct direct examination of company officials, cross-examine other witnesses, put on a rebuttal case, or offer any evidence or other material. Company counsel protested on numerous occasions, but their pleas were consistently turned down.

The lengthy hearings proceeded through such subjects as a decline in employment brought about by expansion of the dial system and the depression's lower business volumes; AT&T's size and its dominance of the industry; lobbying; service to race wires (who obtained almost all of their revenues from illegal bookmakers and practically none from a few token newspaper subscribers); the Bell System pension plan, and whether it discriminated against low-paid employees; Bell dominance in such areas as sound systems for motion pictures; and, late in the game, the rate of return on long distance operations.

Internal dissension at the Commission, although rooted in broadcasting, spilled over into the telephone investigation. Some of the top officials on the investigation staff were controversial, and the heads of the legal and accounting staffs departed before the hearings finished. Carl I. Wheat, considered a more pragmatic advocate of negotiated rate reductions for customers and less of a contentious ideologue than some of his predecessors, was brought in to head staff work in the crucial rates and tariffs area. His separate report on these subjects set the tone for decades of negotiated changes and informal "continuing surveillance" of AT&T earnings, a regulatory style that generated increasing public controversy over the years.

Significantly, general staff work in the final stages of the investigation was put under the direction of Holmes A. Baldridge, who had been with the investigation staff since the early days. When he later left the FCC, Baldridge moved to the Justice Department. A decade later, in 1949, he was lead counsel when Justice filed its first antitrust complaint against the Bell System, principally based on the investigation's findings.

A draft of the staff report circulated among all the FCC members, who clearly were much less than unanimous in approval, before the next shoe dropped. On April 1, 1938, about nine months after the hearings ended, what was variously described as a staff, interim, proposed, or progress report was transmitted to Congress. The view of observers at the time was that the move was intended as a trial balloon to see what the Capitol Hill reaction would be.

If that was true, the effort was a dud. The report was the subject of a lot of newspaper editorial space, with most editorial writers emphasizing the relative paucity of skeletons in the Bell System closet, the quality of U.S. telephone service, and the one-sided nature of the investigation procedures. There was little reaction in Congress and no rush to introduce bills to carry out nineteen legislative recommendations for expanded authority for the FCC.

The report suggested that a 25 percent Bell System local and long distance rate reduction, amounting to about $250 million a year, "may be made." As we have noted, negotiated rate reductions were coming with some frequency in the FCC's jurisdictional area, interstate and foreign service, which probably accounted for some of the crashing uninterest on Capitol Hill. The report proposed, in perhaps the most controversial of its recommendations for change, that the FCC be given power over management policies and decisions in advance of their being carried out. In a country which had apparently survived the depression, Congress was much less likely than in the hectic first hundred days of the New Deal to vote new powers to the expanding bureaucracy.

Although the commissioners spent increasing amounts of time in meetings working on a final investigation report, they were for a while unable to agree completely on much of anything. Chairman Frank R. McNinch had been brought in from the Federal Power Commission by FDR to "straighten out" the FCC, but his poor health and increasing tensions among the broadcast-oriented FCC members handicapped him.

In fact, at the beginning of 1937, FDR had proposed a sweeping government reorganization which included the abolition of *all* of the independent commissions, including the FCC, and the transfer of their functions to cabinet departments. A long Capitol Hill controversy centered on the accompanying proposal to pack the Supreme Court with new justices favorable to the New Deal, and the whole plan finally was shelved.

Two years later, FDR concentrated on the FCC. He called for abolition of the agency and establishment of a new one, expressing "thorough dissatisfaction" with the way things had gone. McNinch, an advocate of the plan, was FDR's official emissary to see it through. Chairman Burton K. Wheeler of the Senate Commerce Committee introduced the administration bill, providing for a three-member Federal Communications and Radio Commission. The committee's senior member Republican, Wallace H. White, Jr., countered with a bill proposing an eleven-member FCC, divided into autonomous broadcast and common carrier panels. Neither plan got anywhere.

In this atmosphere, spiced with some internal FCC disputes which became public, the Commission struggled to come up with a compromise telephone investigation report. It wound up, well over a year after transmittal of the progress report, with a considerably watered-down version. The nineteen legislative recommendations were reduced to nine, and most were of a relatively routine nature. Instead of asking power to make Western Electric, the Bell System's manufacturing arm, a public utility, the FCC requested authority to impose a cost accounting system on Western. There was a suggestion, never taken up by Congress, that the cost of telephone regulation be imposed on the companies. Long gone was the request for power to preapprove Bell System management policies.

Again, there was little public reaction on Capitol Hill. The report was printed as an official document, and seldom heard from again, at least until the filing of the 1949 antitrust suit against AT&T and Western Electric. It did, however, provide the basic informational framework for continuing negotiated rate reductions and other regulatory activities.

McNinch's unhappy tenure on the FCC ended soon thereafter. Suffering from chronic colitis, he was rarely in his office in his final months. His digestion undoubtedly was not improved by events of the last few years, which included various episodes involving George Henry Payne, a maverick FCC member with a journalistic and political background.

At one point, Payne made charges of improper influence on his colleagues by broadcast lobbyists during a hearing before the House Rules Committee. Invited back to a closed session to name names and places, he was reported to have "taken back" the charges and "absolved" the other commissioners. In another series of events unique in FCC history, Payne presided over an investigation of two prominent broadcast lawyers accused of highly unethical acts. When the full Commission took over the case, Payne was disqualified by his colleagues from further part in the matter by a vote of 5 to 1. The lawyers subsequently were disciplined, fairly mildly, by the Commission.

Finally, there were repeated rumors of a "wild party" or "drunken brawl" in a New York night club, with some FCC members attending and broadcast lobbyists picking up the tab. One congressman saw fit to report the rumors in a speech on the House floor. The reports came to a head a year later, in 1940, when Commissioner Thad H. Brown, a member of the FCC and the predecessor radio commission since 1932, was renominated for a new FCC term by FDR.

Finding a vehicle for charges against the Commission, members of the Senate committee held hearings over two and a half months, sometimes for several days at a time, in the usually routine confirmation process. Much of the controversy centered on events leading to a lengthy continuance of a 1932 antitrust trial in which RCA was the defendant. Numerous outside witnesses, including RCA Chairman David Sarnoff, testified in what was supposed to be a hearing on the Brown nomination.

The New York party came in for a few days of discussion. It seemed evident that some sort of social event had in fact taken place, and Brown was present, although a veteran and respected commissioner, T.A.M. Craven, testified he also was on hand and everything was fine. He would have been happy, Craven said, to have taken his wife, mother, or sister to that party.

One casualty of the party, it appeared, was Brown's eyeglasses, broken when a woman who had joined the celebrants slapped him. A casualty of the lengthy hearings was Brown's renomination. After a period of no action by the committee, and with a fight pending on the Senate floor, Brown asked FDR to withdraw the nomination. The FCC term was still vacant the following February when this chapter of FCC history came to an end. Brown was stricken with an intestinal ailment and died after a day's stay in a hospital in Cleveland, where he had returned to practice law.

2

MAKING IT THROUGH THE WAR

Much, although by no means all, of regulation and politics as usual disappeared when the war in Europe began in 1939 and intensified in 1940, and the United States was inexorably drawn closer and closer to the conflict. When the United States finally became a full-fledged belligerent after the Japanese attacks on December 7, 1941, "normal" regulation and politics usually took a back seat to war-related activities, although they never vanished completely.

FCC Chairman James Lawrence Fly, a former Tennessee Valley Authority official and a prominent sub-cabinet member of the Roosevelt administration, was also chairman of the Board of War Communications (BWC). Much of the Commission's wartime staff work was devoted to BWC functions. The board spend much of its time trying to minimize the use of increasingly scarce materials and personnel for civilian-type communications and to concentrate their employment in war-related traffic. For example, it banned social and congratulatory telegrams. Some short-lived political flak flew because President Roosevelt sent Senator Harry Truman a congratulatory telegram when he was nominated as Vice President at the 1944 Democratic convention.

It might be noted, in these days of extinction of message telegrams delivered on the traditional yellow blank, and of domestic U.S. air mail as a separate and slightly higher-priced service, that during the war the telegram and air mail were the principal alternatives to long distance telephone calls.

Telephone service, of course, was not as ubiquitous and affordable as it now is. Wartime telephone facilities were overcrowded most of the time, and little new construction was permitted except to serve military or war-related installations. Consequently, the use of the optional communications means was encouraged.

The importance of the domestic telegraph service was indicated by the attention given the plight of the number two carrier. Postal Telegraph, a part of the ITT system, was near bankruptcy as the 1940s began. It was feared in government circles that its financial collapse would mean the loss of an important service and would disrupt the telegraph business with an adverse impact on the war effort. In 1942, Congress adopted permissive telegraph merger legislation. After hearings, the FCC in 1943 authorized Western Union to take over the Postal Telegraph system.

The general acceptance by policymakers of regulated monopolies was shown again in the terms of the legislation. It authorized acquisition by Western Union of the telegraph services of the nation's telephone companies. However, a principal controversy centered on Western Union's additional role as one of several competing U.S. international telegraph companies. Since its domestic system would now be the only means for the domestic handling of the overseas traffic of all carriers, including itself, Western Union was told to get rid of its international operations. No deadline was set, and divestiture took the better part of two decades. But by the 1980's, Western Union was authorized to reenter the competitive international business. At the same time, policymakers opened all of Western Union's domestic markets to competitive entry.

During the immediate prewar preparedness period, when only staunch isolationists doubted the direction in which the United States was headed, and then after the United States entered the war, one business-as-usual activity which went on was the continuing FCC watchfulness over AT&T's earnings from its interstate long distance telephone service.

There was in the rate and earnings discussions, however, a relatively new and increasingly important consideration. It would outlive the integrated Bell System and become an even more significant factor as the day arrived when the local telephone companies were separate in ownership from the long distance carriers. This was the complex and arcane art (it is difficult to make it a science) of telephone "separations." Basically, it is the allocation of expenses and plant investment between intrastate operations, regulated by the state public service commissions, and interstate service, controlled by the FCC.

Probably nothing is more crucial to all rate regulation. The cost allocations determine the general level of rates that must be charged in each jurisdiction to recover the costs assigned to the telephone operations of that regime. In addition, most long distance calls are handled jointly by two or more companies. Revenues from the joint services have to be divided in some way that will enable each of them to recover its cost of participation. Each company's cost recovery from the total toll revenue pot is determined by separations.

Telephone separations plays a major role in other chapters of this book. It should always be remembered, in considering the subject, that like any other system of allocating costs incurred in producing multiple services or products, separations is largely a judgmental exercise. It was therefore inevitable that jurisdictional separations would reflect political and social factors to no less extent than economic criteria.

As a result, no one separations formula has endured without major change or controversy, since there is always an equally logical-sounding justification for some other way to approach the problem. It is no surprise that separations has been a chronic, nagging source of contention between the FCC and the states for many years.

One other term in the telephone vocabulary is, unhappily, necessary for a discussion of the subject. This is "non-traffic sensitive" (NTS) plant, the great bulk of which is made up of the lines and equipment needed to add a subscriber to the system—whether he or she uses the phone constantly or practically never to originate or receive calls. Its quantity and cost do not vary with the amount of use, or calling, or "traffic."

It consists of the telephone instrument itself, wiring inside the customer's home, the "drop" to the main line outside, a pair of wires in a cable connected to a switching office, and a termination in that office. If a community has 100 homes, each with a telephone, and grows until it has 1,000 telephone-equipped residences, it will have almost precisely ten times as much in NTS plant. This is true even if some of the new homes are vacation hideaways, and if very few calls are ever made from them.

In total, it is a huge item. When the Bell System owned all the telephones and inside wiring in its territory, the amount invested nationwide in non-traffic sensitive plant reached more than $64 billion, or 51 percent of the total which had been put into the ubiquitous telephone network. The problem is that NTS plant is as essential to long distance service as it is to local, but its cost remains unchanged regardless of the number of calls, whether local or long distance, or incoming or outgoing, that traverse the plant.

The reader might now assume the role of a separations negotiator. Let us say the average telephone is in use twenty minutes a day, and 5 percent of those minutes is taken up in interstate long distance calling. Shall we allocate 5 percent of those NTS costs to long distance service? On the other hand, the phone is there, available for service, twenty-four hours a day, equally for local or long distance use. Is 50 percent for each fair? Or should we weight or average actual use of the subscriber line with a factor for its availability? Since interstate calls are obviously over long distances, and cost more with presumed attendant greater value of service, shall we try to arrive at a value of service factor? Or multiply in the average miles of each type of call to arrive at a message-minute-mile factor?

All of these approaches have been seriously proposed, some have been actually used, individually or in combination with others, and all have at least some basis in logic.

Before the advent of serious government regulation of the telephone industry and the emergence of long distance calling as a more commonplace event, jurisdictional separations of subscriber line and other exchange plant costs was considered unnecessary. This was because the charge paid by a subscriber for exchange service also covered his or her incidental access to the toll switchboard for making or receiving long distance calls. Hence, separations dealt only with the jurisdictional allocation of costs on a board-to-board basis. Toll rates were designed to recover only those costs and made no contribution to the upkeep of local exchanges.

In 1930, the U.S. Supreme Court in the landmark case of *Smith v. Illinois Bell Telephone Co.* had ruled that interstate long distance service was not entitled to a free ride on exchange plant and had to take a reasonable allocation of the local costs. However, it took more than another decade before the local ratepayers were relieved of the total exchange cost burden. AT&T took the position that, by the Communications Act of 1934, the Congress excluded the FCC from any jurisdiction over exchange facilities or services, and thus nullified the *Smith* ruling. Many states were ambivalent on the proper separations approach. On the one hand, they were eager for the interstate jurisdiction to pick up part of the local plant costs. On the other hand, they were reluctant to concede to the FCC any jurisdiction over exchange facilities.

As state regulators became increasingly militant and organized, and as the FCC announced one after another negotiated interstate long distance rate reductions, pressures mounted for a change. Why, asked the state officials responsible to their home constituencies for election or appointment, should the long distance users get a free ride on local plant? Why shouldn't the local customer get some of the benefit of increased long distance business and revenues? Since a long distance call was an end-to-end function, why not allocate a share of long distance revenues to help support the cost of local service? In time, this argument prevailed, and the costs of interstate service were determined on a station-to-station basis. ("Station," incidentally, is telephonese for each end of the call, the telephone instrument.) In short, a share of local costs would now be recovered from the revenues of interstate long distance services subject to the federal jurisdiction.

By early 1941, jurisdictional separations had acquired top priority on federal and state regulatory agendas. The FCC had just issued orders launching what appeared would be a sweeping investigation into the reasonableness of AT&T's long lines rates. A month later, filing a statement in response,

AT&T and the Bell operating companies proposed—for the first time, but not the last—that the first step should be an investigation into jurisdictional separations of property and expenses of the telephone carriers between state and interstate operations.

The first result of all this was a "down payment" to take effect in July—a $14 million reduction in interstate rates. The FCC announced that, with the latest rate cut agreed to, the hearings it had scheduled on AT&T Long Lines Department earnings would not be held at that time. The next step was announcement of a cooperative inquiry by the federal agency and the states, through their national association (then the National Association of Railroad and Utilities Commissioners and still today the NARUC), into methods of separations.

(Years later, in an effort to modernize its name, as the once-preeminent job of overseeing railroads faded into near oblivion, NARUC appointed a special committee to come up with a new name. The committee, without ever really explaining except that it wanted to preserve the long-accepted acronym, successfully proposed "National Association of Regulatory Utility Commissioners." It may be the only use of the term "regulatory utility," whatever it means, in the English language.)

A committee of federal and state staff officials—in the states as well as at the FCC, only staff specialists were ready to cope with the complexities of separations—tackled the separations inquiry. At hearings held in Chicago in August 1942, in the relatively early days of World War II, the staff group proposed changes in the process intended to simplify its workings, and based on actual use. The joint or cooperative hearings were presided over by two FCC members—the long-lived Paul Walker, and former California State Commissioner Ray C. Wakefield—and four commissioners chosen from the ranks of the state agencies.

Spokesmen for individual states, at the hearings, split between the board-to-board and station-to-station approaches. The Bell System said it endorsed the staff report, which did not reach formal conclusion on that gut question, and would not offer any additional testimony. The issue was soon to be decided quickly, as a fait accompli, without an official government ruling.

In November 1942, the FCC reported that Long Lines' "excess" earnings were between $47 million and $62 million a year, and that it was making an unheard-of return on net book investment, after taxes, of 14.92 percent. It ordered AT&T to show cause why interstate rates should not be reduced and ordered formal public hearings to start in mid-December.

AT&T took immediate issue with the Commission's earnings calculations, especially those regarding tax expenses, although it was noted later that the rate reduction which eventually resulted was within the FCC's stated range of excess earnings. AT&T's other main point was that it was spending millions in advertising to persuade the public not to make unnecessary long distance calls and to keep the lines open for calls essential to the war effort. A rate reduction, it reminded the FCC, would have the opposite effect.

In its formal answer to the FCC's "show cause" directive, AT&T made the same points, but it also urged that the board-to-board versus station-to-station controversy be settled. As the operator of the interstate system, AT&T registered its support for the board-to-board method. Independent telephone companies, whose cut of long distance revenues was dependent on their investment and call-handling functions in their own local territories, naturally plugged for the station-to-station approach.

After three weeks of negotiations, the FCC announced a $50.7 million Long Lines rate reduction in January 1943. The new rates were carefully crafted to minimize increased public use of the long distance service. Callers would pay the same basic rates for the first three minutes (everyone had a cost-conscious relative who used a three-minute egg timer when making a long distance call, and abruptly hung up when the last sands ran through). Only for overtime minutes above three would callers save an estimated $22.8 million annually. About $12 million more in savings would go to businesses using dedicated private lines.

The remainder of the cutback in Long Lines' earnings, a little over $16 million, was one of those shifts to the local and intrastate side that state commissioners were clamoring for. The "divisions of tolls" received by the local Bell and independent companies would rise that much, supposedly easing the pressure on local rates.

In calculating the increased toll divisions, the negotiators settled the board-to-board versus station-to-station argument. The adjustments were figured on the station-to-station basis. The next month, AT&T and the FCC made it official. AT&T filed, and the FCC accepted, a new tariff definition of message toll service. It said, "The toll service charges specified in this tariff are in payment for all service furnished between the calling and called telephones." This formally resolved a crucial area of disagreement among regulators and telephone companies. But it was also the start of an equally vexing and enduring issue: how to calculate the amount of contribution that interstate long distance revenues should make to the cost of local exchange plant.

Thus began the extremely intimate relationship of interstate rate changes and changes in jurisdictional separation methods, the latter imputing successively larger amounts of costs to interstate long distance operations. For years, when interstate earnings continued at an almost inexorable upward trend, the FCC would be quick to negotiate an interstate rate cut for its constituents. As the trend continued and the states clamored for a share of the interstate prosperity, the FCC would sooner or later accede, grudgingly, to a negotiated change in separations procedures, increasing the amount of common costs allocated to the interstate domain and thus easing the pressure on monthly basic phone rates.

During much of World War II, care was taken not to cut back rates in a way to encourage civilian calling, especially during the peak hours. In July 1943, as Long Lines' earnings continued to move up, the local companies picked up another $20 million a year in toll divisions. Six months later, there was another $8 million rate reduction in interstate services, but about two-thirds of it came from moving up the starting time for lower night rates by an hour, to 6 p.m. local time.

It was not until mid-1944, little more than a year before the war ended, that the government felt confident enough about ultimate victory and the peaking of the U.S. war effort to agree to a rate cut in the basic charges. Another $21 million negotiated rate reduction was announced in the longer hauls beyond 790 miles. The daytime rate for a three-minute transcontinental call went to a new low, $2.50, from $4.

AT&T President Walter S. Gifford protested mildly, noting that Long Lines' rate of return was going up as a result of extraordinary wartime calling volume and the temporary overloading of the system. The FCC proudly countered with the point that the current value of the rate reductions it had been responsible for since it was established ten years before now amounted to $220 million a year.

Wartime business-as-usual activity involving the FCC also centered on the perennial congressional investigations and efforts to reorganize the Commission. As always, the focus was on the agency's broadcasting regulation, with the Executive Committee of the Federal Communications Bar Association leading the way in lobbying.

Senator Wallace White of Maine, continuing as the ranking Republican on the Interstate and Foreign Commerce Committee and probably the member of Congress with the strongest continuing interest in communications, marked the hectic months in 1941 before the nation entered the war by again offering his bill to split the FCC in two autonomous divisions. On the House side, a new member, Rep. Jared Y. Sanders (D., La.), dropped in a bill even more suited to the bar association's views. It not only mirrored Senator White's measure, but also would have directed the FCC to take up and dispose of four main issues which were burning the broadcast industry at the time.

Neither bill got very far, although Rep. Sanders experienced the exhilaration of a first-termer to have extensive committee hearings on a bill with his name on it. It was perhaps fortunate that his first term was exciting, because it also was his last. The next time around, he lost in the primary voting.

Most of the FCC was against the Sanders bill, and Chairman Fly sought to focus the committee's attention on the Commission's common carrier responsibilities. He cited the lack of support for the bill from the common carrier community and argued that its provisions would lead to neglect of the FCC's common carrier duties. In one part of his testimony, he noted that the Bell System's plant valuation at the time was $5 billion, compared to a mere $40 million for the entire broadcasting industry. "You could," the colorful Texan advised the congressmen, "blow the gold dust of the broadcasting industry into the eye of the Bell System and it wouldn't even squint."

Whether the FCC commissioners should be split into specialized panels remained a subject of Capitol Hill argument for some years. In 1947, during a brief one-Congress period of Republican control of the Senate, Senator White acceded to the chairmanship of Senate Interstate Commerce. He again had offered his bill to split the FCC into two divisions, and at his request the FCC filed a draft order taking another approach, to divide the agency's members into three specialized groups.

The FCC plan, which remained under consideration for upwards of two years, called for four-member broadcast, common carrier, and safety and special radio services divisions. The chairman would be an ex-officio member of all three groups, and three of the commissioners would serve on only one panel, as its chairman. The other three would serve on two of the three panels, rounding out their membership.

Commissioner control of the agency's day-by-day work in each of its three main workload areas foundered on internal negotiations about who would get what assignments. The idea's proponents believed that it would make presidential appointees, confirmed by the Senate, responsible for staff priorities and the development of recommended decisions to be acted on formally by the commissioner panels, with review in some instances by the full Commission. But the FCC members, most of whom wanted to be on the broadcast panel, could never agree on the identity of the three commissioners who would be assigned only to the non-broadcast areas.

The concept finally was shot down for good—except for later committees of commissioners who handled some routine telephone and telegraph items—when the Democrats regained control of the Senate during the surprise reelection victory of President Truman in 1948. Put in charge of the Senate Communications Subcommittee was the Senate's new majority leader, Ernest W. McFarland of Arizona, a longtime opponent of the panel system. The subcommittee promptly issued a report critical to the proposal.

Senator McFarland was a believer in FCC reorganization, however, and the FCC fairly quickly read the political winds and put one of his earlier recommendations in effect. An earlier FCC reorganization bill by the Arizonan had called for dividing the staff into three main functional bureaus, the Commission set up a Common Carrier Bureau as the first step in the alignment.

The move was not, as it might have seemed, a mere reshuffling of bureaucratic job sheets. Heretofore, when items came up to the commissioners for consideration, there were often three different recommendations to choose from—those of the law, engineering, and accounting staffs. Now, with all the staffers working on common carrier matters put into one bureau, headed by a single chief, the FCC members were given one staff proposal.

To some extent then, the bureau chief became a power center in shaping regulatory programs and policies. One commissioner nicknamed the staff officials who headed the bureaus "the little kings."

Nevertheless, bureau chiefs who believed that politics is the art of the possible—and they could not rise to the top staff jobs in the agency without accepting that truism—seldom risked repeated rejections of their recommendations by the commissioners. Before sending up their proposals, they usually established that what they were proposing was acceptable to a majority of the commissioners, normally including the chairman. In the common carrier field especially, the bureau chief maintained a particularly strategic posture, since the commissioners usually preferred to rely on the staff to deal with its technical and economic mysteries.

The neat balance between the staff specialists, who spent their working days immersed in the parochial problems of one major line of industry, and the political generalists who comprised the Commission and were more concerned with broader philosophical issues, "looking good," and getting reappointed, turned out to be practical and workable. The bureau chiefs decided, most of the time, what pending issues would get priority, shaped and guided what came out of the staff analyses, and selected what decisional options would be brought to the commissioners.

At the same time, the successful ones kept in close touch with the commissioners, "selling" what actions seemed to be desirable and necessary, and making sure that the often-arcane recommended orders were acceptable as consensus decisions. When the finished product of a ruling in a controversial case emerged from the Commission, it was usually impossible to tell whose opinions had predominated—the staff who had done the writing and analysis, or the commissioners whose names were on it as having voted for it.

That the setup achieved a bureaucratic balance rare in Washington is clear from its longevity. The bureau form of organization has survived for more than thirty-five years, and the collegial Commission as a source of debate over setting specific day-to-day policies has declined. Chances for extensive input from commissioners with varied backgrounds diminished still further in the early 1980's. Congress decided that the best way out of a political dilemma which arose was to reduce the number of commissioners to five from seven, allowing the terms of two incumbents, a Democrat and a Republican, to expire. No voices urging direct commissioner control of the FCC staff's daily workload were heard.

3

THE FIRST CHINKS IN THE DIKE

Much regulation as usual may have been "on hold" during World War II, but the lines buzzed with some of the forces that helped shape postwar telephony. The all-out national effort to win the fighting, fueled with all available resources and almost unlimited spending, included unprecedented activities in basic and applied research and development.

Scientific and engineering search into the radio spectrum, keyed to radar and electronic surveillance, greatly accelerated the advent of microwave radio relay. The new type of intercity communications had been in the research stage at the Bell Telephone Laboratories before the war, but the intesified projects of the war brought it to the point where AT&T could file applications for such an experimental facility with the FCC by March 1944.

AT&T proposed a commercial service trial of microwave relay between New York and Boston, and an authorization was issued later in the year. When the war ended, construction began apace over heavy-traffic routes, aimed at a transcontinental system.

Commercial tests of coaxial cable, which also provided the ability to carry several hundred conversations or a television program over a single conductor, had begun before the war. Now, construction of both coaxial and microwave systems, complementing each other in a single national network, moved rapidly. The two types of systems, with their heavy capacity and economics of scale, revolutionized the economics of long distance telephone calling and the burgeoning new industry, television.

Television stations were going on the air in many cities across America. Their viewers, at first able to watch network program material only on films and kinescopes, hungered for "live" TV from outside their own communities. AT&T public relations people kept up a running scorecard as the TV service network steadily expanded, linking new points to the network program centers weekly.

Six years after the end of World War II, AT&T held its own "golden spike" ceremony, with microwave and coaxial "legs" now linked across the continent. TV viewers in cities all across the nation were able to see the Atlantic and Pacific oceans simultaneously, in one dramatic telecast, for the first time.

In terms of economics and ease of installation, microwave systems eclipsed coaxial cable in most areas. They soon became the principal medium of the intercity network. Because the towers were spaced about forty miles apart, on relatively high ground so that the antennas would have line of sight with each other, land use rights were needed only where the towers were built. The high points of mountainous terrain, which would have been impassable for the plows laying coaxial cable, were made to order for microwave radio relay.

The microwave systems, in tandem with the coaxial cables, changed the national concept of long distance telecommunications from an expensive wonder to the readily accepted commonplace.

The most visible example was television. Viewers quickly became accustomed to rapid switching of news and special events from one origination point to another, hardly pausing to consider the wonder of it all. In a relatively short time, as history is measured, viewers would accept as equally routine and ordinary their ability to sit at home and watch events taking place across the oceans, brought to them by satellite transmission.

But, while TV was the national telephone network's showcase, its workshop was the rapidly growing use of long distance calling. Long-haul telecommunications' contribution to the gross national product moved up steadily, in percentage and amount, as the public progressively made greater use of it. The high-volume long distance systems meant lower costs of transmitting intercity calls, and lower costs meant lower rates.

While improvements were being found in the wartime laboratories, two things which were *not* happening in the telephone industry would have considerable effect on the early postwar scene. Under wartime restrictions, telephone companies were installing few new residence telephones, and the piles of "held orders" for service were mounting. When the millions of servicemen were discharged—large numbers of them establishing new households—telephone companies joined other producers of goods and services in meeting heavy pent-up demand for the things that were rationed or in tight supply during the war years.

It was nearly a decade before subscribers in all parts of the country could get telephone service promptly after placing an order. The Bell System reported in mid-1946 that it had installed 1.7 million telephones in the first six months of the year, but that the backlog of held orders on July 1 exceeded that amount and stood at 1.8 million. The backlog was fairly well cleaned up until the Korean War of the early 1950s again delayed the efforts to wipe it out completely.

Wartime conditions also led to considerable labor ferment after hostilities ended. Strikes were generally frowned on during the war as unpatriotic, and employees became accustomed to bigger paychecks as a result of overtime and the upward pressure on wages stemming from shortages of qualified workers. Veterans came out of the service with higher expectations, and union leaders believed after V-J Day that the time had come for them to act aggressively to gain benefits deferred during the war.

The whole situation added up to a series of major strikes, and the Bell System was no exception. Unionism in the telephone industry had moved from company-sponsored employee associations to a rather loose confederation known as the National Federation of Telephone Workers. NFTW leaders had as one of their goals in a six-week strike, in April and May 1947, the establishment of nationwide bargaining, pitting the national union against the national Bell System.

The united front lasted for about a month, but began to break apart as some of the individual groups started reaching agreements with their employers. While nationwide bargaining did not succeed in 1947, the effort finally transformed the NFTW into "one national union"—with some exceptions as some of the state or companywide groups remained unaffiliated—the Communications Workers of America.

Another wartime development which was to affect postwar telephone service and regulation was telephone recording. It created some of the first exceptions to the accepted underlying premise that good service was dependent on the telephone companies providing all the equipment attached to the line.

As observed earlier in this book and as it will be again in the next chapter, the accepted cornerstone of U.S. telephone service was the totality of AT&T control of the system and the service, carried out through the vertically integrated structure of the Bell System. Organizationally, the Bell System did the research, made the equipment, distributed it, installed the facilities, provided the service over them, maintained them, and handled the accounts and money for the entire system. It divided the revenues among the participating companies, including the independently owned companies whose local exchanges were interconnected to the Bell network.

Because the telephone companies were thus accountable for end-to-end service, in all of its ramifications, the accepted truism was that they must provide all the equipment that was used in furnishing it. Only in this way, telephone management and regulators agreed, could the companies maintain the control necessary to fulfill their responsibilities.

The point was well illustrated in what may be the first case of its kind in which someone with a new product or idea brought a complaint to the FCC because the Bell System would not accept it. An inventor named Cammen had developed what he regarded as a better telegraph printer, and he tried unsuccessfully to sell it to the Bell System for private line telegraph and teletypewriter exchange service use.

In a 1936 ruling on his formal complaint, the FCC turned him back on all counts. The record, the Commission said, "does not warrant a finding that [Cammen] has, or may be reasonably expected to have, a printer better than or as good as printers now used on [AT&T's] lines; or that [Cammen's] proposed printer, if and when perfected, may be used on lines of [AT&T] without interfering with other services which [AT&T] holds itself out to furnish the public; . . . or that the Commission should, under the showing made, set out objective rules and standards which would determine in advance the suitability of a machine without the necessity of a test under service conditions."

Under the exigencies of wartime, telephone recording became a fairly widespread practice in World War II. This was particularly true in the military and the government, for several reasons.

One was the obvious desirability in a wartime atmosphere of precise records of what had been said in giving instructions and orders or reporting on emergency situations. Another was the inherent caution of military and other bureaucrats who did not wish to be blamed if things went wrong, and might be able to establish by a recording that the error "didn't happen on my watch." Finally, things could be done under emergency conditions which would not have been allowed in normal times.

Bell System companies would never give any public acquiescence to invasions of the privacy of telephone service. The mere suggestion that one party to a call might make a recording for some later disclosure was enough to send shudders down the spine of the collective Bell System management. That kind of "listening in" to a private conversation received about the same kind of reaction as any suggestion of indiscriminate wiretapping. On the other hand, party line listening in was tolerated as an economic and service necessity.

Nonetheless, as a wartime measure, under the prodding of the Board of War Communications, AT&T revised its federal and state tariffs to allow electronic recording in government offices. It would have been difficult under war conditions to refuse the insistent demands of the Army and Navy that recorders were urgently needed as a military measure. Besides, the military presumably would have done it anyway, tariff or no tariff. Permission to use recorders in government offices was made an explicit tariff exception to the general ban on "foreign attachments"—foreign, in this case, was anything not provided by the telephone company.

Even the wartime measure came under the article of faith of the time: If something was connected to a telephone line, it had to be done by the telephone company, which would also provide and maintain the connecting equipment. No one argued against it, or even considered any other possible arrangement, short of the telephone company becoming the manufacturer and supplier of the recorders themselves, which it chose not to do.

Total acceptance of the prohibition against foreign attachments, and the ban on interconnection of anyone else's products or property, went far beyond the concern about privacy of telephone conversations. Not only to telephone company management but to the federal and state officials who watched over the system as regulators, it was simply unthinkable that an outside party would be permitted to tamper with the telephone line, let alone interpose some gadget into the sacrosanct telephone system. It did not make any difference who was doing it, or why; it just would not be done.

A story from the early 1930s, when Herbert Hoover was still president, exemplifies the accepted and inflexible rule. The teller was a prominent Washington attorney who in the early days of his career was the chief staff officer of the then-fledgling National Association of Broadcasters. Let the late Phil Loucks pick up the story:

We were a brand-new organization, and holding our first real convention in Detroit. President Hoover was a big radio buff, and was responsible for the creation of the Federal Radio Commission when he was Secretary of Commerce. My big coup was arranging at the White House for the President to address the convention.

In those days of train travel, he wasn't about to make the trip to Detroit. But the solution seemed very simple. He would simply pick up the telephone in his office—that in itself was fairly new—and call the NAB in Detroit. The assembled broadcasters, in the meeting room, would hear his remarks over an amplified public address system.

Everything was great, until we called Michigan Bell while setting up the convention arrangements. We told their installers what we wanted, and they flatly refused—it was a violation of the interconnection rules to connect a telephone to a public address system. Nothing I could say made any difference to them. I was getting desperate, and started calling top officials of Michigan Bell. I got the same story from each of them; it didn't make any difference if it was the President, and if no harm was being done to anyone. It was a violation of their tariffs.

As the time approached for the presidential call, not knowing what else to do, I called AT&T President Walter Gifford at his New York office. I was told he was travelling somewhere in Tennessee, and when I explained the situation, they said they would try to get back to me.

When time was really getting short, I got a call from Gifford. I poured out my problem, and he laughed and said, "Well, in an organization as big as this one, we have to have rules. Leave it with me, and I'll see what I can do." That took care of it, and the interconnection was made. We got the call, and the broadcasters heard the message. But wouldn't you know—President Hoover was tied up on urgent business, and some clerk in the White House read his message to us.

Now that the war was over, and their national emergency exception permitting them to use recorders was imperiled, government departments, led by the Navy, urged the FCC to look into the subject and come up with some new foreign attachment rules. In a fairly unusual procedure at the FCC, which usually starts investigations after some petitioner asks for action, the Commission held a two-day fact-finding hearing on the possible attachment of recorders in January 1946.

Commissioners' questions of witnesses reflected the Bell System's concerns about privacy of conversations. They heard testimony from Bell and independent telephone companies, the NARUC as the representative of the state commissions, recording device manufacturers, large users of telecommunications service, and the ubiquitous—on this subject, at least—Navy Department.

A warning was sounded that telephone customers who wanted to record conversations were going to do so anyway, regardless of what the tariffs on file at the regulatory bodies said. Witnesses for the Dictaphone and Sound-Scriber companies said they had sold more than 15,000 recorders between 1937 and 1945. The Bell companies reported that in the first eleven months of 1945, they had received orders to install only forty-one recorders.

Eight months later, the FCC all but decided the case, in the context of the belief of the times that telephone service had to be protected against harm from outside devices. The Commission prescribed the protective connecting devices or arrangements which later became so controversial after the Carterfone decision, and in dozens of antitrust cases. It issued a proposed report concluding—tentatively, it was said—that recording devices had legitimate and valuable uses for commercial and government purposes. But, the FCC went on, they would have to be connected with adequate service safeguards.

In addition, the FCC said, users would have to be informed that their conversations were being recorded. It suggested two possible methods of alerting subscribers: an automatic tone warning signal, and a special indicator or asterisk in telephone directories, showing that the listed telephone was equipped with a recording device. The Bell System companies came up with another suggestion, under which a caller about to use a recorder would have to make the call through the operator, who would announce to the other party that the device was on the line.

An "important precaution" to be taken against unauthorized use, the FCC said, would be the requirement for physical connection. It turned down the proposal of recorder manufacturers that acoustic and inductive coupling be used (nothing would be plugged in). Again reflecting the accepted wisdom of the day, the FCC stated, "All of this connecting equipment, as distinguished from the recording apparatus itself, should be provided, installed, and maintained by the telephone company."

The Commission heard oral argument a few months later, but did not finally decide the case for well over a year, near the end of 1947. It stayed close to its previous rulings, once again insisting on a connecting device furnished and maintained by the telephone company. It laid down technical standards for the connections, including the requirement that users be alerted by a tone warning signal (known universally as the "beep tone") about every fifteen seconds, equal in strength to the signal of the average telephone conversation.

The agency had issued its first general rule permitting connection of foreign attachments, but how widely it was complied with never was really known. The beep tone was considered something of an annoyance from the start, with complaints that parts of the conversation were being obscured by the warning signals.

In time, radio and television broadcasters, airing telephoned reports from correspondents dispatched to scenes for stories, took particular issue with the interference caused to those transmissions. Presumably, since the broadcasters' use of the recordings was public, they felt more compelled than some other subscribers to comply with the rules. Ultimately, the Commission made an exception for them.

General belief was that considerable recording was being done surreptitiously for private purposes and that the very use of the beep tone probably defeated a lot of those purposes. For the most part, recorded telephone conversations were not admissible as evidence in court cases. The FCC asked for information on enforcement of the tariffs in the mid-1960s and was told that in the entire Bell System during the preceding year a total of 234 cases of possible tariff violations had been investigated. Recorders had been found to be connected according to the rules in forty-one of those instances. Most of the other subscribers who were investigated promised to discontinue recording telephone conversations, or said their machines never had been used to record them.

Much later, with the march toward deregulation in Washington, and with divestiture only a few months away, the FCC came close to throwing in the sponge in the fall of 1983. It put out a proposed rulemaking notice citing the evident enforcement problems and questioning the legality of the existing rules, since the recorders were about the only devices still requiring a telephone company-furnished connecting arrangement. As will be recounted in later chapters, this was a requirement that the FCC had already scrapped for all other classes of foreign attachments as well as for customer-provided telephone equipment of any kind, such as the telephone instrument itself. The agency saw sound reasons for discontinuing the regulations altogether.

Only two weeks before divestiture, AT&T effectively ended the matter, although it still refused to concede publicly that the privacy of telephone conversations was being invaded. For the first time, the tariffs officially permitted the use of acoustically and inductively linked recorders. They also formally provided an alternative to the customer alert contained in the beep tone, this being prior mutual consent to having the recordings made. The consent, AT&T, dying hard, told its customers, should be in writing or made part of the recording if oral.

With few exceptions, the commissioners went into the 1950s almost totally engrossed with and overwhelmed by the rise of an exciting new industry, television. Center of attention, and the real bonanza for Washington communications lawyers, focused on the contests for television licenses in almost every significant market area. Public hearings attracting by far the most attention were keyed to proposals for TV broadcasting in color.

When pressed, and without an in-depth examination of the philosophy, the FCC stuck closely with the long-accepted basic principles of common carrier regulation. They were a dim view of new entrants, as long as the existing facilities were adequate, and the opinion that good telephone service was best insured if the telephone companies controlled the equipment it took to furnish it.

Nevertheless, in one other relatively early hearing case, the FCC made a small break in the dike of monopoly. It did so mostly by accepting a neat way out offered by the hearing examiner who conducted the evidentiary part of the case.

This involved telephone answering and recording devices, the now familiar devices which answer an unattended telephone and accept a message from the caller to be played back by the telephone subscriber on his or her return. For one-person offices, as well as occupants of residences not wishing to miss calls, the need for this type of service was evident. Producers of rudimentary and not always reliable early models, known as the Telemaster, Telemagnet, and Electronic Secretary, besieged the Commission with pleas to be exempted from the foreign attachment ban. In mid-1950, the FCC ordered a hearing on the complaints and soon thereafter broadened it to include a general investigation of the subject.

The stakes here were somewhat different than in other FCC cases. It was arguable whether telephone service was at issue, since the called telephone was ringing when it activated the somewhat clumsy early answering devices to answer, in some cases by actually picking up the telephone handset. The problem was how well the devices worked. FCC staff members tested them and reported their findings at a hearing held in early 1951.

At the same hearings, AT&T reported that it had decided it was feasible to provide a grade of commercial answering and recording service which would be acceptable to subscribers. The Bell companies would offer to lease the instruments to their customers. Selected was a device known as the Peatrophone, and 200 units had been ordered by AT&T for extensive testing. The FCC staffers reported that their own tests had shown that the best performance, in a series of statistical tests, came from the Peatrophone. The others were found to be less reliable.

By the middle of 1951, the case had reached the point of proposed findings and conclusions by the parties to the hearing. The FCC staff still stuck to the approach which the Bell System later, to its peril in private antitrust cases, adopted as the one that had government blessing. The staff said that unrestricted use of the devices could have adverse effects on telephone service. Consequently, the staff drew upon the precedent of the earlier 1948 recording device ruling. The Bell companies should be told to file tariffs permitting use of the equipment "if such devices are connected to the telephone line by means of connectors or isolation units which assure protection to the service and facilities of the telephone companies."

AT&T backed off a little, giving the FCC a cue it later accepted. It could see "some demand" for answering and recording devices in exchange and intrastate toll service, regulated by the states, but a "negligible" need for answering interstate calls.

By November of the same year, examiner Basil P. Cooper—not a notable student of the law, but someone who knew a solution when it was offered to him—had his answer. His initial decision concluded that the state regulatory commissioners were in the best position to decide whether attachment of a particular device to a specific type of telephone would cause problems with the service.

Cooper held that the foreign attachment provisions of the interstate tariffs filed with the FCC "should not be used to prohibit the installation and use by the subscriber of a telephone answering device not furnished by the telephone company in any community or state in which the use of such answering device is authorized by appropriate local or state regulatory agencies or commissions."

The answering device sellers had been trying to win their point on the local level, too, although several states had rejected similar complaints by the producers of the Telemagnet against corresponding state tariff prohibitions. But, Cooper noted, "The record reflects no impairment of interstate and foreign toll service of the Wisconsin Telephone Co. as a result of the Public Service Commission of Wisconsin requiring [the company] to permit subscribers to continue the use of Electronic Secretary." He observed that "with rare exceptions, telephone answering devices when installed will answer intrastate and local exchange calls more often than they will answer interstate and foreign toll calls."

The following February, in the customary comments on such initial rulings, the FCC Common Carrier Bureau stuck to its guns. Among other things, it warned the FCC that it would be abandoning its jurisdiction if it decided the states were in the best position to regulate answering and recording devices. On the other hand, AT&T argued that the FCC did not have jurisdiction and pointed out that Bell companies had now made tariff offerings of answering devices in twenty-nine states and the District of Columbia.

Still deep in TV issues, the Commission did not assign any pressing priority to the dispute. It was almost two years after an oral argument on Cooper's initial decision before the FCC, in May 1954, came out with a final decision. The delay appeared to have resulted in part by a split in the ranks of the commissioners over the outcome.

Finally, by a 4 to 2 vote, the Commission announced that it did not foresee enough use of answering and recording service in connection with interstate and foreign calls to upset the Bell System's tariff ban. But, it declared, it would not stand in the way of states or localities authorizing the use of the devices. Thus, customers were given another small opportunity to attach their own equipment to telephone lines without obstruction from the FCC but, of course, provided that the distributors could win the approval of regulators in their state. Significantly, three of the four FCC members comprising the majority were former state commissioners.

The states generally took a dim view of the early devices, joining those who considered them somewhat clumsy and unreliable, and equipment marketed under the brand names of the complainants to the FCC in the early 1950s has disappeared from the electronic store shelves. But the inventors and developers continued their work, hoping either to sell the products to telephone companies offering the devices to their customers or to individuals who did not spend their business days perusing telephone company tariffs. Events following the Carterfone decision gave them a broader government-blessed market, and ultimately deregulation of terminals, or customer premises equipment, threw the field wide open.

In the interim, another set of entrepreneurs found a different way to satisfy the needs of businessmen and others who did not wish their sometimes-unattended telephones to miss calls from potential customers, employers, or friends. The telephone answering service sprang up, with live operators responding to calls switched to a central location by the equivalent of off-premise extensions. The appearance in modern electronic telephone central offices of "call forwarding" service—incoming calls are switched to another telephone number designated by the customer whenever the user wishes—gave the answering services another shot in the arm.

Telephone answering, like most activities connected with U.S. telephone service, is big business. After divestiture, the answering services' trade association, Associated Telephone Answering Exchanges, reported that its members had 1.2 million customers and were answering 860 million calls a year. There was no way to estimate how many telephone subscribers were using their own answering devices.

4

THE COURTS TAKE A HAND

Certain truths about the telephone industry were held to be inalienable in the middle of the twentieth century. These went beyond the simple concept that telephone service had to be a regulated monopoly to afford consumers maximum service at the lowest possible rates.

One truth was that the telephone company was wholly responsible for the service from one end to the other. This went from the caller picking up his handset, getting a dial tone, and signaling the network what number he wanted, to the ring at any one of millions of telephones at the called end and, if the called party answered, throughout the transmission of their conversation until the caller hung up and the disconnection told the vast telephone network that the call was completed.

To fulfill that responsibility, it was carved in marble in the minds of those whose concerns about telephone service went beyond picking up a phone and expecting it to work that the telephone company had to have complete control over the system. For anyone else to substitute a piece of equipment in that system was unthinkable (a belief which, as we will see, went even beyond the landmark Carterfone decision of the FCC).

Equally unthinkable was the use of "foreign attachments" to telephone company property. Uninhibited use of any sort of customer-provided gadget ran the risk of interfering with the service, since the telephone company could hardly guard against harm by controlling the design and production of the device and its maintenance upon need or demand, unless it was the supplier.

In those rare instances where it seemed there was a good public policy reason to permit limited use of foreign attachments, as with automatic answering machines and conversation recording devices, elaborate precautions had to be taken to prevent harm to the service. It was not enough to assume that users of foreign devices were consenting adults who were hurting only themselves. After all, users could call any one of those millions at the other end, and their service, as well, must be protected.

Less frequently mentioned, but clearly understood to the government and industry policymakers, were two considerations. First came what amounted to the cherished right of private property. The nation's telephone companies had spent billions installing a pervasive nationwide system available to anyone who could afford a few dollars a month. Any private entrepreneur could make a fortune if he or she could sell any sort of gimmick to be attached to even a small fraction of those millions of telephones. But since the telephone companies' stockholders owned the system, why should entrepreneurs be allowed to do that?

Second, the entire base of the telephone industry was built on recurring monthly charges for what amounted to leasing, rather than selling anything for a cash price to consumers. The telephone companies would argue that they were not just leasing but providing facilities to furnish a service, not a jumble of leased equipment. The fact remained that all of those dimes and quarters on monthly telephone bills for lights, shoulder rests, and all sorts of telephone-company-provided attachments helped pay the freight. To regulators, if the telephone companies had enough gadgets out there—leased or provided or whatever—the revenue would help keep down Aunt Minnie's basic monthly telephone bill. Cash for the sale of a piece of foreign equipment going into the pockets of store owners didn't do a thing for Aunt Minnie.

The first break in this line of thinking, now totally a thing of the past, came from the inventor and proprietor of an obscure device known as the Hush-A-Phone. The business of Harry A. Tuttle, he ultimately told the Commission when his case finally came to hearing in early 1950, had been severely inhibited by telephone companies' reminders to retailers, and to some consumers, about the "foreign attachment" provisions of the companies' tariffs.

The tariff provision, quite simply, was that anyone using a foreign attachment was subject to discontinuance of telephone service.

As a rule, except in the rare occasions when a formal proceeding or investigation is held and the regulatory agency officially prescribes rates or tariff language, the carriers file their tariffs with the regulators to become effective on some specified future date. The tariffs, unless suspended or rejected, go into effect on the scheduled date.

It is usually a misnomer to say that a tariff has been approved by the Commission. Typically, it is just allowed to become effective. Nevertheless, once effective, the tariff has the force of law.

The foreign attachment rules had been in effect for a long time. Although a few minor exceptions had been made, they had never been subject to serious challenge as a rule of general application. Not, that is, until Tuttle came along.

As far as was known, no one's telephone service had ever been discontinued because he or she used a Hush-A-Phone. But the threat of a possible cutoff was raised often enough to dissuade prospective buyers, and store owners often discontinued handling the product when advised by phone company representatives of the possible perils to their customers.

Enforcement of the foreign attachment rules was usually sporadic. A telephone installer or repairer who found a foreign device on a premises visit usually would simply disconnect it, sometimes with a word to the subscriber. If a telephone company manager saw a store advertising some device which fit the description, the manager might visit the store and raise an admonitory word.

The classic case usually cited was the successful effort of a telephone company to halt distribution of plastic decorator covers for telephone directories, carrying the advertising of the distributor. The directory, it was contended, was telephone company property, and the cover was an attachment. AT&T officials, asked about the incident, said that such nitpicking was not Bell System policy, and that the action was taken by an overzealous local manager.

To the casual observer, the Hush-A-Phone hardly looked like a serious threat to the integrity of the nation's telephone service. It was a cup-shaped plastic device which fit over the mouthpiece of the telephone handset. The other end fit over the telephone user's mouth. If a businessman, say a real estate salesman, wished to converse privately with the seller of a home while a prospective buyer was sitting in his office, he could use the Hush-A-Phone. He could also, it was pointed out later by the FCC, obtain the same effect by cupping his hand over the mouthpiece. Installation of Hush-A-Phones in offices where a number of employees were talking with customers, it was observed by its developer, would reduce ambient noise and cut down interference with each employee's ability to hear what was going on.

Tuttle was obviously not one to rush to the government for help at the first sign of business trouble. He had been making the Hush-A-Phone since 1921, or nearly three decades before his complaint finally led to an FCC hearing. The FCC apparently saw nothing landmark in the case. The Commission issued an initial decision, subject to objections, legal briefs, and oral argument, a full year after the hearings ended. The agency's final decision—which turned out not to be so final after the trip to court—was not issued for almost another five years, at the close of 1955.

The Hush-A-Phone entrepreneur said he had sold about 125,000 units of the device since 1921. Of those sales, 84,000 were of the model designed to fit the ancient deskstand, or pedestal, telephone, which had been gradually replaced over the years with handset instruments, starting in 1927. (For more than fifteen years, some telephone companies imposed a surcharge, usually 10 cents a month, for handsets.) The Hush-A-Phone was sold for about $10 retail or $6 wholesale to department stores.

When the complaint finally moved to formal hearings, Tuttle and his counsel—Kelley Griffith, who later went back to government service and will be prominent in a later chapter—proved themselves pioneers in another respect. They brought in two college faculty members to serve as expert consulting witnesses. Dr. Leo Beranek of the Massachusetts Institute of Technology was a specialist in acoustics, and Dr. Joseph C. P. Licklider of Harvard was a lecturer in psychology.

They testified that while the Hush-A-Phone did in fact reduce somewhat the level of the talking voice in the desirable range of frequencies, it improved the intelligibility of conversations where one of the parties was seeking privacy. In terms of the ratio of words overheard by an eavesdropper when the Hush-A-Phone was used, as compared with the caller whispering or cupping a hand over the telephone, the parties did better with a Hush-A-Phone, they testified. They also reported that the device was effective in reducing outside noise levels.

Tuttle recommended a change in the foreign attachment provision. It would permit the user to attach his own device, if it were not connected electrically, and if the FCC did not find that it adversely affected telephone service.

Bell System witnesses argued that the transmission and receiving losses resulting from use of the Hush-A-Phone would nullify all the improvements in instrument design over several decades. They were, they reminded the FCC, responsible for good service in the eyes of the public and its designated guardians, the regulators. They pointed out that foreign attachment provisions in telephone company tariffs went back to 1913 and were included in customer contracts as far back as 1899.

In its initial decision, the FCC stuck by its traditional guns. In its crucial conclusion, a good example of a bureaucratic sentence but one that is understandable if read carefully, the Commission said, "Where a device has a direct effect upon communication itself, as does the Hush-A-Phone, if we were disposed to do so at all we would require a showing far stronger than that made by Hush-A-Phone herein to warrant departure from the general principle that telephone equipment should be supplied by and under the control of the carrier itself."

After the further proceedings dawdled along for years, a unanimous Commission finally dismissed the Hush-A-Phone complaint. First, the agency gave its blessing to the existing foreign attachment tariff rule. It commented, "Inasmuch as the unrestricted use of foreign attachments by the public may result in an impairment to the quality and efficiency of telephone service, damage to telephone plant and facilities, or injury to telephone company personnel, it is necessary and proper that the use of foreign attachments be subject to control by [the telephone companies] through reasonable tariff regulations."

The Commission conceded, "It appears that the use of the Hush-A-Phone device affords some measure of privacy as well as a more quiet telephone wire by reason of exclusion of surrounding noise, and that no physical damage of any consequence results in telephone facilities when the Hush-A-Phone is used."

But, the FCC went on, "It is clear from the record that the use of the Hush-A-Phone for the primary purpose for which it was designed [privacy] is accompanied by an impairment in the quality of telephone transmission. As shown by our findings of fact, the use of the Hush-A-Phone results in a number of effects upon the telephone circuit: a slight loss of intelligibility when used in ordinary conversation, and a greater loss of intelligibility when used with the objective of obtaining privacy."

Following the customary approach of the time, the FCC took the position that if anyone wanted a telephone line free from outside noise, the user could obtain it by getting equipment which was readily available from the telephone companies. It observed that "telephone users may obtain from the [telephone] companies 'push-to-listen' and 'push-to-talk' switches which may be used to exclude noise from circuits."

The bottom line, the FCC concluded, was that "the significant factor here is the Hush-A-Phone's feature of providing privacy to a talker in a quiet environment. This benefit, however, does not balance the public detriment involved in the loss which . . . is caused in intelligibility, together with other adverse effects which . . . result from the use of the Hush-A-Phones."

As far as the FCC was concerned, that was the last word. But Tuttle had not given up. Two months later, in a move which was unusual in those days but is commonplace to the point of routine today, Hush-A-Phone asked the U.S. Court of Appeals for the District of Columbia Circuit to reverse the FCC and remand the case to the Commission. Hush-A-Phone charged that the FCC failed to consider that "the antitrust laws are violated by the tariffs." It also told the court that the Commission was remiss in failing to rule that the Bell System was "unlawfully interfering with the natural and inherent rights of a subscriber to use an office device related to the efficient conduct of his business."

It was in mid-November of the same year, 1956, that the court, after the usual legal briefs and arguments, handed down its decision. The unanimous three-judge panel ruled, in words which later were cited to open the telephone system to much wider use of customer-provided equipment, that the foreign attachment provisions "are unwarranted interference with the telephone subscriber's right to use his telephone in ways which are privately beneficial without being publicly detrimental."

The court concentrated much of its opinion on the effects, legal and practical, of subcribers in essence creating their own Hush-A-Phones by cupping their hands over the mouthpiece and speaking softly.

Heart of its opinion was the following dissertation:

Although the Commission found that using a Hush-A-Phone does not physically impair any of the facilities of the telephone companies, it nevertheless concluded that the device is "deleterious to the telephone system and injures the service rendered by it." There seems in that conclusion a suggestion that the use of the Hush-A-Phone affects more than the conversation of the user—that its influence pervades in some fashion, the whole telephone system. . . . It is because we see no findings to support these conclusions of systematic or public injury that we reverse the Commission's decision . . .

The question, in final analysis, is whether the Commission possesses enough control over the subscriber's use of his telephone to authorize the telephone company to prevent him from conversing in comparatively low and distorted tones.

It would seem that, although the Commissioin has no such control in general, there is asserted a right to prevent the subscriber from achieving such tones by the aid of a device [the Hush-A-Phone] other than his own body. . . . Thus intervenors [the telephone companies] do not challenge the subcriber's right to seek privacy. They say only that he should achieve it by cupping his hand between the transmitter and his mouth and speaking in a low voice into this makeshift muffler. This substitute, we note, is not less likely to impair intelligibility than the Hush-A-Phone itself . . .

In both instances, the party at the other end of the line hears a comparatively muted and distorted tone because the subscriber has chosen to use his telephone in a way that minimizes the risk of being overheard. In neither case is anyone other than the two parties to the conversation affected.

To say that a telephone subscriber may produce the result in question by cupping his hand and speaking into it, but may not do so by using a device which leaves his hand free to write or do whatever he wishes, is neither just nor reasonable.

The federal appellate court having spoken, the parties were free to take the case one step higher, to the U.S. Supreme Court, if they chose. But the FCC and AT&T decided to treat the court decision as "no big deal." Instead, they set to work preparing revised tariffs which would reflect the court's directive, but in as narrow a construction of it as possible.

Tuttle's suggestion at the hearing for a revised foreign attachment rule was accepted almost without change. The new language, submitted by the telephone companies and accepted by the Commission after a public notice asking comments produced little response, gave subcribers the right to attach "foreign" equipment which suited their convenience, provided that it did not endanger telephone employees, property, or service. To continue to protect the telephone system and its employees, it could not be connected electrically, inductively, or otherwise.

A few months later, in September 1957, the Commission dismissed the now-satisfied Hush-A-Phone complaint.

The direct results of all this were not remarkable. Hush-A-Phones did not sprout from the nation's telephones, and, today, except for its name on a case which produced one memorable sentence, the device is virtually forgotten.

The narrow interpretation of the court's ruling, translated into tariff language, did not open the door wide for all kinds of gadgets. It did take a few attachments, such as shoulder rests and directory covers, clearly out of the "foreign attachment" classification, where they had been—in terms of the purpose and concept of the rule—all along.

Nevertheless, the Hush-A-Phone case built the threshold for the Carterfone decision, the FCC program of registering terminal equipment to certify that it was safe and not harmful, and today's "all comers" subscriber equipment market.

As those developments moved through the governmental process two decades later, no one forgot that telephone subcribers had a federal court-ordained right to use their telephone "in ways which are privately beneficial without being publicly detrimental."

> *Strassburg's comment:* I served throughout the Hush-A-Phone proceedings as the FCC's staff counsel and recommended against authorizing use of the device. In my view, Tuttle's prospects of winning his case before the FCC were, from the start and without court intervention, virtually nil. This may be somewhat difficult to believe in the context of current FCC policies feverishly promoting deregulation, competition, and maximum free choice for customers.
>
> But the Hush-A-Phone case is a classic illustration of the regulatory values that dominated the entire telephone regulatory community for generations. They were embraced by the FCC from its beginnings in 1934 and into the late 1960s. Thus, it was the conviction of the FCC and its staff that they shared with the telephone company a common responsibility for efficient and economic public telephone service and that this responsibility could only be discharged by the carrier's control of all facilities that made up the network supplying that service. Such control included not only transmission, switching, and the subscriber station used for basic end-to-end service. It also had to extend to any equipment a subscriber might attach to or interface with the basic service. Only by this comprehensive type of control could the quality, safety, and economies of network performance and design be assured.
>
> Hence, the blanket ban on a customer's use of foreign attachments without specific authorization was accepted by the regulator as defensible.

The Hush-A-Phone, by itself, posed no threat of any real consequence to the performance of the basic network. Nevertheless, authorization of its use could set a precedent for other, less benign attachments which individually or cumulatively could degrade network performance to the detriment of the public.

This was the litany of the regulator that was Tuttle's impossible burden to overcome at the FCC. (It is interesting to note that the FCC with a Democratic majority in a proposed decision in 1951 unanimously voted against Tuttle, and that some four years later, with identical rationale, a Commission with a Republican majority likewise voted unanimously to deny use of the Hush-A-Phone.)

Fortunately for Tuttle and the future of regulatory policymaking, the court took a more balanced view of the Hush-A-Phone and its like. In effect, the court's ruling provided a rational reconciliation between the rights of subscribers on the one hand and the limits of telephone company responsibilities in the protection of their networks on the other hand.

Those of us who, on the regulatory scene at the time, could be classified as political liberals and distrustful of monopolies, were not entirely comfortable with our support of the foreign attachment restrictions. Clearly, they expanded the scope of the Bell System's monopoly beyond its basic network. At the same time, they fortified the manufacturing monopoly maintained by AT&T through Western Electric of all facilities used in Bell System services. AT&T's tariff restrictions, together with its manufacturing monopoly, obviously stood as an effective disincentive for any independent, innovative, competitive development of communications gear that might be used in connection with basic telephone services. The result, we know, was to leave the United States industrially unprepared to respond effectively to the domestic penetration of foreign-manufactured communication products as FCC policies turned to competition for the supply of telephone equipment and services.

But we liberals could rationalize that the public policy or antitrust questions raised by AT&T's exclusionary procurement policies were separate from or subordinate to the overriding conviction that protection of the basic network service from any actual or potential degradation had first priority. In addition, we took illusory comfort from the pending government antitrust suit begun in 1949 to force the divestiture of AT&T's manufacturing monopoly and thereby open the Bell System's captive market to a competitive supply of equipment. That suit was, of course, aborted. But, more on that later.

5

THE SURVEILLANCE CONTINUES

The FCC's first serious postwar try at formal public investigation of Bell System interstate rates created such a political firestorm that it was nearly fifteen years before it was repeated, in a dramatically different environment.

Immediate reaction of industry officials to the Commission's bombshell announcement of a show-cause proceeding in January 1951 was one of shock. Influential public officials, led by state public service commissioners, and soon thereafter the majority leader of the Senate, joined them. Underlying it all was the complexity of the regulatory establishment under the separation of state and interstate jurisdictions in the Constitution's commerce clause. Responsible for telephone regulation within their borders, the states saw in the FCC's action a move to benefit interstate users of a joint enterprise to the detriment of the customers of services limited to state boundaries.

For several years before the FCC made its seemingly sudden announcement, the earnings of the AT&T Long Lines Department, derived entirely from interstate services, had been running at relatively high levels. FCC staff memoranda to the commissioners calculated those earnings to be 10.01 percent in 1946, 6.89 percent in 1947, 6.99 percent in 1948, 6.3 percent in 1949, and 7.6 percent in 1950. During that period of relatively low cost of money, 6 percent was something of a standard for regulated utility earnings, and most regulators did not quickly become alarmed about a company's ability to raise capital if earnings dipped for a while closer to 5 percent. AT&T's own calculations of its rate of return were somewhat lower than the FCC staff's, since it espoused a rate base larger than the Commission's "bare bones" favorite, net book cost of plant.

There were several reasons why the Bell System was doing so well in the early postwar years. Telephone volume was booming, with interstate revenues and messages (calls) usually setting new records each year. High-cost long-haul wire and cable circuits were being replaced with efficient modern facilities. Pressure for substantially higher wages was just beginning to be felt, and the next inflationary upsurge as a result of the Korean War was still a little distance away. New construction programs, bringing with them large demands for new capital, were heavy, but had not yet reached the massive size that they did a little later in the postwar era.

There was another reason, to become progressively more important during the next decade, for comparatively high Bell interstate earnings. The methods for dividing revenues, expenses, and investment between the interstate and intrastate local services, or telephone separations, seemed to favor the interstate or Long Lines operations. This was the complaint of the state commissions which provoked the series of subsequent changes in separations methods, each of which shifted some investment and costs, or revenue requirements, from the intrastate to the interstate or FCC jurisdiction.

The first separations manual, specifying the general formula for making the jurisdictional divisions, had gone into effect in 1947. The product of several years' effort by a committee of federal and state commission staff members, it was aimed at simplifying the allocations methodology, rather than being any kind of philosophical or policy revision. It embodied the premise reported earlier that all long distance service was end-to-end, or station-to-station. To reflect this principle, the new procedures moved a relatively modest $13 million in annual revenue requirements (expenses plus the carrying cost of the investment) from exchange operations to the interstate jurisdiction.

Rumblings during 1950 that another big interstate rate reduction might be in the offing began to arouse state commissioners, who were busy at home coping with the first postwar wave of local and intrastate rate increases. The Bell companies, looking at somewhat higher wages and prices in the initial postwar inflationary push, were deluging state regulators with rate increase applications. They argued with some success that local rates, and toll charges within state borders, many of which had not been increased in twenty years or more, just were not enough to cover the higher expenses they were facing.

The states were particularly reluctant to raise basic monthly rates for service, especially of the residential variety. They often "found" at least some of the money the local telephone companies needed by raising intrastate toll rates, believed to be paid predominantly by business and more affluent residential customers.

In most of the states, this led quickly to aggravated toll rate disparity, the condition under which a toll call within state boundaries cost appreciably more than an interstate call of the same length. During this period, a staff study under the auspices of the NARUC, the state regulators' association, showed that intrastate calls were costing $125 million more a year than the same number of calls, over the same distances, between points in different states and thus charged for under the interstate rate schedule.

Something else was bothering the state commissioners as they gloomily surveyed their local situations, complete with newspapers carrying screaming headlines about telephone rate applications in their states. Almost all the cost-cutting produced by research was going to long-haul transmission facilities. Little had been done to improve the cost efficiency of local telephone systems, or of the short-haul toll systems which carried mostly intrastate traffic.

The way the situation looked, local and intrastate telephone rates had nowhere to go but up, while long-haul interstate charges were going down.

Even though the FCC made no public overtures toward a rate reduction, AT&T and the states could read the signs and the numbers. AT&T had a practical political problem and understandable economic concerns. Its low earnings in most states meant that it was necessary to file the rate increase proposals, but their size endangered their acceptance and strained relations with the regulatory commissioners. Meanwhile, the economies of scale in the long-haul system were pushing down unit costs, leading to higher interstate earnings and an inevitable interstate rate reduction and making things that much worse for the states.

Although it was the regulators who adopted separations changes, AT&T was not—nor could it have afforded to be—a passive onlooker. Since it owned the entire interstate system, and most of the intrastate operations, its financial interest in allocations of costs between the two was obvious. In all of the separations changes which came in the 1950s and 1960s, AT&T was, as it was later observed in a background memorandum by Strassburg to the Justice Department in preparation for the second antitrust suit against AT&T, "a principal, if not controlling participant, in the process." The memo went on:

> While the regulators may have initiated each effort and even conceptualized the approach to be taken, they were in no position to convert a concept or objective into a specific plan or methodology, or to estimate the results of its application on the revenue requirements of the several jurisdictions. This is because the regulators lacked adequate expertise and the essential data exclusively in possession of AT&T. For these reasons, the regulators were forced to look to AT&T to structure any proposed formulation and assess its results on the revenue requirements of interstate operations and intrastate operations in each of the states.

The first of a series of new separations proposals to take its name from the location of an NARUC meeting where it was endorsed gained its formal approval at the 1950 NARUC convention in Phoenix. For the states and AT&T, the so-called Phoenix plan was an ingenious concept whose time had come. It did not arise from any studied analysis of separation principles, but rather from the mutual conviction of the state regulators and the telephone company that an FCC move on interstate earnings was near.

It was founded on the fact that the cost of toll plant owned by the Bell associated companies (the local telephone companies) was at least twice the cost per circuit mile of the longer haul toll plant of the Long Lines Department. Long Lines plant was used exclusively for interstate calls, and therefore its book costs were assigned entirely to that jurisdiction. The toll plant of the local companies, on the other hand, carried both intrastate and interstate calls, and those costs thus required a jurisdictional allocation. The Phoenix plan would have treated the Long Lines plant in each state as if it were jointly used for state and interstate calling, by combining its low costs with the higher cost toll plant of the local telephone company and then allocating the aggregate on a relative use basis. By this process of averaging, low-cost long-haul circuits which simply passed through a state would be used to pull down the investment assigned to state operations, and the difference would be assigned to the interstate jurisdiction.

To the FCC, cost separations had been based on relative use, at least since a 1930 Supreme Court decision in *Smith v. Illinois Bell*. It felt that the accident of geography by which long distance toll circuits merely passed through a state on their way to a distant terminal did not justify what amounted to a subsidy of the telephone operations of the state through which they passed.

The convening state regulators were told by FCC Vice Chairman Paul A. Walker, still a charter member of the federal agency who was coming to the end of a career including the FCC's special telephone investigation of the 1930s, that the FCC considered the Phoenix plan "unsound and objectionable." Commissioner Walker was a long-time member of the NARUC Executive Committee, but that did not stop the convention from endorsing the Phoenix plan without audible dissent.

Walker, in his speech, opened the door to a cooperative federal/state effort to make mutually acceptable separations changes. But, to him and others, the time was coming to tackle the problem head-on. Perhaps lacking was an understanding that the situation had grown larger than the FCC.

During the next two months, conventional wisdom outside the FCC was that the agency would take no action on long distance rates until it had taken a good look at the NARUC request to test the Phoenix plan. The plan, it had been calculated, would move $20 million in expenses and $200 million of toll plant investment to the interstate jurisdiction. The effect would be to relieve the states of some $40 million of revenue requirements that would otherwise have to be defrayed in local and intrastate rates.

Without advance notice, late on a Friday afternoon January 22, 1951, the FCC ordered an investigation of the Bell System's interstate rates (and foreign rates, but they were a small part of the whole). At the same moment, FCC and state commission staffers were meeting to wind up a year-long study of the causes for the toll rate disparity problem as a basis for seeking a solution. Despite *Telecommunications Reports'* occupation with keeping up on what was happening, it described the FCC's action as the "best-kept regulatory secret of recent years." The Commission ordered the Bell companies to show cause why their rates should not be reduced and scheduled hearings to open April 16.

Hopping mad, the states found their champion, and in Washington he was not an inconsiderable one. Amiable Ernest W. McFarland of Arizona may not go down as one of the Senate's immortals, but he had important credentials. A long-time member of the Senate Commerce Committee, he had repeatedly shown a strong interest in communications matters. And he was so highly regarded by his colleagues in the "world's most exclusive club" that he had been elected Majority Leader by the Democrats when that influential post was vacated by the incumbent, who had been unexpectedly retired by the voters in his home state. In one of history's fortuitous accidents, the states could not have found a better time for the chairman of the Senate's Communications Subcommittee to be Majority Leader of the Senate.

McFarland responded to the challenge. He wrote the FCC a strong letter, saying in essence that this was no time to consider further interstate rate reductions when there was a "fantastic" disparity between state and interstate telephone rates. The difference, he pointed out, was running at more than $100 million a year. The FCC, he told the office-holders two miles down Pennsylvania Avenue whose reappointments were subject to Senate confirmation, should give the green light to the Phoenix plan.

Thirty-six years later, some of Senator McFarland's words were still being heard on Capitol Hill, as some members of Congress objected to the monthly subscriber line charge imposed by the FCC to help pay for the customers' access to long distance service.

"The trouble is," Senator McFarland told the FCC, "the general public does not realize that every move that is made to reduce long distance toll rates results directly or indirectly in an eventual increase in local exchange telephone rates and in intrastate toll telephone rates. But very plainly and simply, this merely shifts the load from the big user to the little user; from the large national corporations which are heavy users of long distance to the average housewife and business or professional man who do not indulge in a great deal of long distance but are the life blood of the telephone business in this country."

The FCC could not buy the Phoenix plan, considering it a substitution of fictional costs for actual costs. But it understood the political realities of the situation. In less than four weeks after its announcement of the big rate investigation, it had postponed for four months the dates by which AT&T was to show cause why its rates should not be reduced, and the hearings were to start. This, the FCC said, would give it time to study separations changes and observe the trend in revenues and expenses.

Negotiations went on during much of the year, and in mid-October, the regulators had their answer. At NARUC's 1951 convention in Charleston, S.C., Walker presented what quickly became known, appropriately enough, as the "Charleston plan." It had been preceded by further postponements of the FCC's rate investigation, and by the report on state/interstate toll rate disparities by a state-federal staff subcommittee. That group reported that intrastate schedules were higher than interstate in all but two states, and that intrastate toll schedules were 35 percent higher than those for points between the states.

The Charleston plan, which applied to the jurisdictional allocation of exchange rather than toll plant, was devised by AT&T, with the encouragement of FCC officialdom. It was said to be designed principally to simplify jurisdictional separations without deviating from established principles. But, in the process of simplification, the use of large-scale averaging of costs was substituted for costs which formerly had been directly identified and assigned. It was by this averaging that the large majority of states reaped some benefits.

The new plan, the now-happier state commissioners were told, would move $90 million of local exchange plant investment and $22 million of associated annual expenses from the states' jurisdiction to the interstate side. In terms of annual revenue requirements, the transfer totaled $30 million.

"Charleston" was the first direct attack on what continues to be a serious stumbling block in allocating long distance telephone costs—the extremely high proportion of local subscriber plant investment to the total. Even now, when the telephone instruments themselves are deregulated and no longer telephone company property, the cost of subscriber lines terminating in the local telephone central offices is a huge amount. Since some 25 percent of the cost is now allocated to the interstate jurisdiction, it amounts to the tail wagging the dog of total long distance service.

To gain an appreciation of the amount of capital required just to connect each subscriber to the central office, look at the figures for the Bell System in 1980, the last full year of end-to-end telephone service. Total Bell System plant in service at the close of that year was more than $125 billion. Of that, subscriber plant consisted of more than $64 billion, or about 51 percent. Station equipment, mostly telephones, comprised 20 percent of the total;

access lines amounted to 27 percent; and terminations into the central offices provided the other 4 percent of the 51 percent.

For allocation purposes, the Charleston plan split local exchange plant into
four major classes, compared with the previous sixteen. Each was then
allocated between state and interstate on the basis of state versus interstate
minutes of use. That meant more averaging and less direct assignment to a
jurisdiction. It greatly simplified the separations procedures while paying
lip service to the FCC's relative use yardstick, but it eased the burden on the
states a little.

By splitting exchange plant into four major categories, instead of the sixteen existing before the Charleston plan, the plan considerably enlarged the
amount of investment classified as subscriber line or access. It is almost impossible to quantify the amount. But it is obvious that substantial costs that
would be subject to direct assignment, or treated as traffic sensitive, were
not to be treated as part of subscriber line NTS (non-traffic sensitive) plant
whose combined costs would be allocated on a relative use basis.

The states felt better, but they wanted more. Especially, they still wanted
a piece of the benefits of those high-volume, long-haul interstate systems
with their economies of scale and low cost per circuit mile. Their staff experts again looked long and hard at the Phoenix plan to find a way to make
it acceptable to the FCC.

Soon, they encountered a new problem. Under the combined load of appreciably higher wages and capital costs, as well as the impact of the new
separations procedures, interstate earnings were going down. The interstate
cow as not dry, but it was showing at least temporary reluctance to supply
all concerned with almost everything they wanted.

All separations proposals advanced by the states had the precise effect of
an interstate rate reduction—a cutback in the interstate rate of return.
When AT&T's interstate earnings were relatively high, it had an obvious incentive to cooperate and assist in making separations changes. But times
had changed.

Even though total estimated revenues in 1952 were 11.6 percent higher
than the previous year, AT&T's earnings were down to 5.6 percent using the
FCC staff's smaller rate base, and about 5.2 percent on the company's net
investment base. (AT&T's rate base included cash working capital and construction work in progress, two items of long dispute between regulators
and the company.) What was worse, the trend was downward, and AT&T
estimated that its return was about 4.5 percent for the second half of 1952.

Despite its position that a formal rate investigation might result in findings that would raise AT&T's return figures by disallowing some parts of
the rate base, the FCC Common Carrier Bureau was ready to agree, at least,
that the earnings were not unreasonably high. At the beginning of 1953, it
recommended, and the Commission agreed, that AT&T's wartime
"employment stabilization reserve" should be released.

The reserve was the product of a deal made by the FCC with AT&T in 1944, during the latter stages of World War II, in lieu of a rate reduction which it was feared would stimulate demand and overtax already over-burdened long distance plant. The company was given approval to set aside some of its excess earnings in a reserve account to pay for future mainten-ance, deferred where possible during the war, and to counter anticipated postwar unemployment. By the end of the war, the fund totaled $11 million. (It was estimated that, because of high wartime corporate income taxes and the excess profits tax, it took before-tax income six times that much to pro-duce the $11 million.)

Once the war was over, it was hard to identify specific maintenance pro-jects which had been safe to defer. Also, postwar revenues were sufficient to pay for current and deferred maintenance with no adverse impact on the bottom line. With maintenance returned to its usual level, the FCC told the company to keep the fund segregated until further orders. It decided that the money would be used to benefit ratepayers, by avoiding or delaying future rate increases which otherwise might be found necessary.

With an economic slump occurring in the early 1950s, the time had come to agree to AT&T's renewed request that the employment stabilization reserve be released. The company was told it could use $1 million a month during 1953 for general corporate purposes until the fund was all released. The action would raise the company's rate of return for the year by about six-tenths of a percentage point.

That was not enough to help much. By September 1953, the FCC staff advised the commissioners that for a test year ended June 30, AT&T's earn-ings were 4.5 percent on its rate base and about 4.8 percent by the staff's calculation. Either one, the commissioners were told, represented a "defi-ciency in interstate earnings" as was "insufficient to meet the minimum cost of capital of the Bell System."

After informal discussions in which AT&T obtained the sense of the FCC, it filed new tariffs in late August looking toward a $65 million rate in-crease. At today's traffic volume, a rate increase of similar proportions would produce additional revenues more that ten times that amount. Volume of the business had grown to the point where this amount could be obtained without any boost in basic daytime initial period rates. The money would come from 5-cent increases in overtime minutes, above the three-minute initial periods, on almost all calls, and 5- and 10-cent rises in the basic rates for most night and Sunday calls.

It was estimated that the higher rates would produce a 6.5 percent return on the FCC staff's rate base and 6 percent on the company's. Not raising any objections, but submitting the analysis "without recommendation," the bureau's memorandum was a model of caution.

Weighing the options, the Common Carrier Bureau's Acting Chief, Cur-tis M. Bushnell (Bushnell had been 'acting' for a couple of years and never was promoted to a full-fledged chieftaincy) gave the FCC members both sides. He said that to the extent that a "comprehensive investigation" of matters in dispute "would result in the disallowance of items of claimed investment or expenses," the outcome would be a cutback in the rate increase. On the other hand, he noted, "If a formal investigation were instituted, the Bell System would probably contend for substantially greater additional revenue requirements" than the amount at stake.

Clearly, conventional regulatory standards dictated that the FCC suspend the proposed rates and set them for formal investigaton and hearing. They represented a general increase of substantial magnitude—the first of its kind in FCC history.

The staff identified a number of aspects of AT&T's claimed revenue requirements that were subject to question. The FCC had never examined these questions in a formal investigation. Major rate changes had frequently been made in the past without formal proceedings. However, those changes generally involved rate reductions and not increases. To permit the current increases to take effect with no formal action of any kind meant that the FCC had no question concerning their justification or was resolving any questions in favor of AT&T.

The FCC, fairly promptly, went along with the AT&T proposal, letting the tariffs go into effect. As the next two chapters make clear, the FCC's inaction put it on the defensive for years to come.

Lines of the differences over the Commission's action were laid out in separate statements by the one dissenting commissioner, Frieda B. Hennock, a Democrat and the FCC's first woman member, and in a reply by Commissioner John C. Doerfer, who was solidly with the majority. Doerfer was a Wisconsin Republican who had chaired his state's public utility commission. (Doerfer had been slated for appointment to the Federal Power Commission, but a mix-up of papers in the Eisenhower administration inadvertently named him to the FCC.)

Commissioner Hennock pointed out, "This is the first general interstate rate increase in the history of this Commission. In passing upon the reasonableness of such a sizeable rate increase without sworn testimony and evidence subject to cross-examination, I believe that the Commission is abdicating its responsibilities under the Communications Act. . . . Moreover, a hearing in this case is particularly necessary because the Commission has never established, on the basis of a public record, the fundamental principles and policies by which to judge the reasonableness of Bell System interstate rates."

Doerfer replied that her philosophy "disapproves a rate increase without a full public hearing but fails to disagree with rate reductions without a full public hearing. The [FCC] since 1940 has effected reductions of over

$250,000,000 without a public hearing. It is my view that the call of duty requires prompt disposition of rate matters both ways.''

The next chapter was close to inevitable. In March 1954, the order to start a full-fledged rate investigation had been on the FCC's books for more than three years. The water over the dam during that period had included the drop in AT&T earnings, as well as the separations changes. Interstate earnings were lower than forecast when the $65 million rate increase was allowed. To make things worse, the rate of return was still trending downward.

All told, the staff recommended that the thing to do was to terminate, officially, what almost had been the FCC's first big telephone rate investigation. The commissioners went along, and the dormant fuse of the January 1951 bombshell was quietly pulled.

Thereafter, things began to improve a bit, however slowly. Federal and state staff members were engrossed in the study of possible separations changes keyed to the interexchange, or long distance, part of the business. AT&T, still unhappy with its earnings, was not joining the project with any special enthusiasm.

Regulators on both sides of state boundaries finally arrived at a variation of the Phoenix plan they could agree to. In typical fashion, it became known as the "modified Phoenix plan" and the NARUC endorsed it at its 1954 convention in Chicago. It had the informal acquiescence of the FCC. AT&T did not oppose it philosophically, but in the words of an active participant, it "dragged its feet." Its endorsement of the plan, AT&T made clear, was contingent on an interstate rate increase to offset the cutback effect of a separations change.

For the next year, off-and-on closed negotiating sessions were held, and meanwhile AT&T's interstate earnings began to improve. By January 1956, the FCC was able to advise the NARUC that it had agreed to put the plan into effect, as usual on an "interim" basis. In a letter to New York Commissioner Spencer B. Eddy, who headed the NARUC Communications Problems Committee, the FCC noted that its action was in no way tied to an interstate rate revision.

"Modified Phoenix" had much in common with the original Phoenix plan, which the FCC had rejected as unsound allocation and unlawful interjurisdictional subsidy. Nevertheless, the revised system passed muster with the Commission. Unlike the earlier plan, it included in the cost-averaging process only those Long Lines interstate circuits which terminated in the state where the plant costs were being averaged. This was endorsed on the rationale that these circuits, although used exclusively for interstate calls, served the same users in common with the local company's high-cost toll plant, which carried both state and interstate calls. Long Lines plant that merely transited the state was excluded from the averaging process.

Thus, once again, direct assignment was departed from. As with the original Phoenix proposal, averaging in the long-haul facilities brought down the cost of plant serving the state. As the plan went into effect, it trimmed intrastate revenue requirements, moving them to the interstate side, by $40 million a year.

Until 1965, when the FCC not only ordered a full-fledged rate inquiry but actually carried it out, the watchword remained "continuing surveillance." FCC members always argued that the process was more effective than carrying on time- and money-consuming rate investigations, plodding through lengthy and complex sessions, finally reaching a decision, and then very possibly finding that while the formal record was being made, conditions had changed and they had to start all over again. Rate reductions could be achieved more quickly with continuing surveillance.

With continuing surveillance, it was contended, the FCC stayed on top of the situation all the time. The staff analyzed detailed financial and operating reports, some submitted as frequently as monthly, while talking with company officials regularly and getting answers to their questions. The information thus obtained, "continuing surveillance" proponents felt, was more reliable than what would have been gained through formal, adversarial investigations, with witnesses constantly on guard in public sessions and attorneys developing records with one eye on possible later court appeals.

Periodically, when interstate earnings began to move beyond the "bounds of reasonableness," FCC and AT&T representatives would hold more formal sessions. Company officials and members of the financial community brought in by AT&T would be questioned by the FCC commissioners and staff. They were, however, entirely closed to the public and other interested parties.

In continuing the surveillance process, the FCC and AT&T had two things going for them. First, FCC regulation of long distance message telephone rates and AT&T's overall earnings ratios was considered a wholly two-party process. There were no consumer advocates worthy of the name. Large users who could afford lawyers and economists were interested primarily in specialized services like private line and telpak. For the most part, these services were not included in the negotiation of rate reductions or increases, thus neutralizing any threatened intervention of the large user.

Incredible as it seems today, no one protested effectively. The Commission was not forced to open up the process, as any agency which attempted it today undoubtedly would be required. When AT&T filed its $65 million rate increase in 1953 after a round of talks with the Commission, no more than a half dozen letters of objection were received by the FCC, none from national or broad-based organizations.

The second reason probably accounted in large part for the first. The process worked pretty well. We have noted rate reductions which, cumulatively, had an enormous effect on the amounts the public paid for telephone service. Increases rarely took place, and then only when the viability of the "world's best telephone service" was in some perceived danger. The cost of an interstate telephone call might not have been slashed to the last nickel or dime wished by ardent populists, but, on the whole, telephone customers were doing all right.

What was wrong with the process, in the view of many professional regulators, was that some nagging unresolved questions were never answered. First, no attempt was made to probe beyond the figures presented by AT&T in order to ferret out expenditures representing waste or inefficiency.

Second, the FCC did not fix, in an overall sense, an authorized rate of return for AT&T until it held its first full-scale rate case in the mid-1960s. The same circumstance applied to some disputed items in the rate base. Third, it could be said that for all the years the Bell System held together, until the January 1, 1984 divestiture, the propriety of Western Electric Company prices and profits on sales to the operating companies was never fully resolved. Finally, AT&T's capital structure, relatively heavy on high-cost equity and light on lower-cost debt, was not tested until the 1965 rate case.

One limited opportunity had been found earlier to seek answers to some of those issues, when AT&T filed its multiple-channel, private line tariff in 1955. Part of the impetus for that tariff was AT&T's service to the big SAGE aircraft defense warning system. Government officials felt that the volume of its circuit requirements warranted some sort of cheaper-by-the-dozen reduction.

Underlying philosophy of the multichannel tariff was similar to that of the later telpak, to be recounted in chapter 8. A per-channel unit measurement system was applied, and several voice channels over the same route would be entitled to the lower multichannel rate. Rates for eligible channels also would decline with distance. Multiple channels over 250 miles would be priced at 90 percent of the regular per-mile rate, and at 1,500 miles the mileage charge would drop to 60 percent.

Western Union, still AT&T's main domestic competitor, filed a protest on grounds of unfair competition, and the FCC suspended the multichannel tariff and ordered a hearing. At the start of the hearing, a detailed cost study of the various private line services was directed, and the sessions were postponed to allow the study to be made. AT&T, legally entitled to put the tariff into effect once the ninety-day maximum suspension period was over, agreed to postpone it for a "reasonable" period until a decision was made.

At the behest of the Common Carrier Bureau, the FCC broadened the multichannel case into an all-out investigation of common carrier—AT&T and Western Union—private line services. Issues included a proper rate of return for the services, the contested rate base items, and Western Electric prices.

Because of the time needed for preparation, especially the cost study, the hearings did not begin until January 1958. Soon thereafter, the General Services Administration, acting for the government agencies as users of communications services, filed a motion requesting that private line voice channel rates be cut 25 percent. GSA did not offer any independent analysis of its own but relied entirely on AT&T's direct presentation, including the results of AT&T's cost study. The FCC in June ordered an immediate 15 percent reduction in the rates for private line telephone channels. It said this would cut AT&T's rate of return on the service to 7.5 percent. There was no indication that GSA had done much except ask for the rate reduction, but its team working in the case received substantial cash awards for exceptional service to the taxpayers.

In July 1961, after lengthy public hearings, the FCC issued an initial decision in the case. Essentially, it reduced private line voice grade rates, the province of AT&T, and raised private line telegraph charges, where the always hard-pressed Western Union provided some competition. The decision, the FCC said, would mean a 7.25 percent return for AT&T in providing the private line services. The agency concluded that the riskier Western Union was entitled to 9 percent, but found that the AT&T rates provided an effective competitive ceiling beyond which the charges could not go.

At the same time, the Commission made its first ruling on some of the disputed rate base items. It excluded plant under construction, cash working capital, materials and supplies, and the plant acquisition adjustment account, all of which had been argued for inclusion by AT&T. Property held for future use was excluded in part. In the process, the FCC rejected the multiple-channel tariff.

AT&T had presented its usual case on Western Electric prices, emphasizing the fact that Western sold equipment to the telephone companies at lower prices than were available from the smaller independent suppliers. While in essence rejecting the case presented, the FCC made no adjustment for Western prices, concluding that the evidence was not sufficient for any action either way. At the same time, it wrote a letter to Western urging "serious and immediate consideration" of a price reduction.

The FCC had a small staff working on such matters, and no one pushed all that hard for action in common carrier matters unless major political forces were on its tail. So it was a year and a half before the Commission issued its final decision in the private line case. It was remarkably similar to the initial decision. An important element of the whole proceeding was that the rate changes were prescribed by the FCC, based almost exclusively on cost of service.

Upshot of the rate decision was that telegraph rates were higher, in part because of the elimination of "clock hour," or time-of-day, discounts, permitting customers who could take advantage of them to get lower rates in off-peak hours. The Commission found that in private line service, there was no cost difference, and that rates should be determined independently of the time at which service was furnished.

The private line case was notable for several things. First, it relied entirely on fully distributed or historical costs by which to judge the level of earnings on individual classes of services. Second, it was the first formal proceeding in which the FCC tackled some of the long-disputed classic rate issues. Finally, as has now become routine in any case in which important economic interests are represented, it was appealed to the courts. Three meat-packing companies did not like the telegraph portions of the ruling and took it to the U.S. Court of Appeals in Chicago.

The court upheld the Commission up and down the line, in a decision issued in what became known as the "Wilson case," named for the first of the three appealing companies, in August 1964. It accepted the FCC's cost-based rate approach and the fact that the Commission's rate-of-return findings were based largely on cost of capital considerations.

In the interim, the case also gave the Commission some defense against congressional criticism that it did not formally investigate AT&T rates. In mid-1958, when Chairman Emanuel Celler of the House Judiciary Committee, whose criticisms are reported in the next chapter, wrote to the FCC on the subject, the agency was able to reply that the private line case was affording it "an opportunity to review general ratemaking principles and policies in the telephone field."

Arrival on the scene in 1961 of new FCC Chairman Newton N. Minow, appointed by the incoming President Kennedy, brought a few changes in continuing surveillance which moved it closer to the modern mode. For one, the FCC staff was allowed to hire a few outside expert consultants to counter the economists and financial analysts that AT&T usually brought along. One was Dr. James C. Bonbright, a widely recognized authority in the field. Bonbright, author of a number of textbooks, consistently came close to the middle of the road in a trade in which almost all practitioners are indelibly labeled "for" or "against" the utility, and usually run true to form.

Minow, a Chicago attorney who had been an associate of Adlai Stevenson, was enough of a pragmatist not to try to tamper with the basic system of continuing surveillance.

Continuing surveillance conferences always had been closed, and the transcripts regarded as confidential. This time, however, the word was quietly passed that the transcripts would be shown to those legitimately interested.

> *Henck's comment:* Probably no better illustration of how times have changed exists than my experience when the first transcripts were made available. It was an important story, and I played it up pretty big. Later, when a competing publication did what was obviously a rewrite of my story, I checked with the custodian of the transcripts. I had been the only reporter who asked to borrow them. Today—in the highly unlikely event that 'continuing surveillance' could be used at all—the news media would be knocking down the doors.

The first such sessions in the Minow regime led to the "after 9" plan, in which there was a transcontinental maximum of $1 for a three-minute station-to-station call anywhere in the forty-eight contiguous states between 9 p.m. and 4:30 a.m. The revenue reduction of $55 million was partially offset by $25 million in increases for interstate calls most expensive to handle—operator-handled calls up to 800 miles—for a net reduction in consumers' bills of $30 million. The increases in rates for shorter distance also made the toll rate schedules of the states look better, since they lessened the amount of the disparity at those distances between state and interstate long distance schedules.

Another series of continuing surveillance sessions a year and a half later, starting in July 1964, resulted in what was, up to that time, the largest telephone rate reduction in history. Of the $100 million total, three-quarters came from moving up the "after 9" $1 transcontinental top rate an hour to the more convenient 8 p.m., as well as applying it all day Sunday. A short time later, the other $25 million was produced by reducing rates for daytime station-to-station calls (customer dialing was limited then, and a station call was simply one destined for a telephone number, rather than a specified person) over 600 miles. Now, one could make a three-minute station call anywhere in the contiguous United States for $2.

Nevertheless, the pressures for a formal rate investigation continued to mount. It was not long thereafter that the Commission, apparently reacting to increasing criticism of continuing surveillance, took the plunge and ordered what turned out to be its first full-fledged interstate rate and earnings investigation (see chapter 10).

The states remained the biggest critics of the process by which, in their view, the FCC was siphoning off most of the cream from the business and not returning enough to the states in the form of separations changes. They were not completely satisfied later in 1965, before the FCC's formal rate hearing was ordered, when negotiators meeting in Denver came up with another revision, moving $134 million in revenue requirements to the interstate jurisdiction.

No others went as far, however, as the California Public Utilities Commission. Under what turned out later to be a correct assumption that the FCC would necessarily deal with separations if it held a full rate investigation, the state, in the early part of 1965, took the negotiated $100 million interstate rate reduction to the U.S. Court of Appeals in San Francisco. It did not oppose the interstate rate cuts, but rather called for a formal record. The California agency emphasized that intrastate customers were entitled to "their share" of rate reductions.

It quickly developed, when the case was argued in January 1966, that the state did not have a chance in the federal tribunal. This became obvious even though Dan Ohlbaum, arguing for the FCC, said the Commission did not wish to have the case dismissed as moot because it had ordered a rate investigation in the interim. Seldom in the experience of longtime observers had appellate judges made their ruling so evident by their questions and comments from the bench during an argument.

The expected denial of the California request came quickly, less than a month after the argument. In its opinion, the court relied on basic common carrier law. It said the rates were "carrier-made" and not "commission-made," and thus, since the FCC had not ordered or prescribed them but merely let them go into effect, this exercise of discretion by the FCC was not subject to review of the courts.

Before the court case, and the FCC's order for a formal rate case, closed-door separations negotiations had another round. After the $100 million interstate rate cut of early 1965 was successfully behind it, the FCC suggested to the NARUC a more intensive program of cooperative discussions on separations. By the end of July, the talks had produced the "Denver plan."

The Denver plan ended the fourteen-year-old "Charleston assumption" that there was no difference in the book costs per minute of use of subscriber line plant whether used for local or long distance calls. The Denver plan was based on a rather complex arithmetical formula—although not as complex as the later "Ozark plan," which remained in effect largely unchanged for more than a decade.

It was applied to the "one for each line" non-traffic sensitive plant, making the arbitrary assumption that the local loops had greater worth for minutes of long distance use than they did for minutes of local use. Two elements of the formula, the number of users during the month and the minutes of use, in effect were double-counted on the interstate side. The result was that a minute of interstate use was worth more than twice as much in terms of separating the costs of the subscriber plant between the state and federal jurisdiction.

It quickly developed that, based on more up-to-date information than had been available when it was negotiated, the Denver plan would transfer considerably more to the interstate side than the original estimate. It also tended to run afoul of a long-term FCC precept that separations should not be employed simply to make arbitrary shifts of costs to orchestrate rate changes.

FCC Chairman Dean Burch made this observation four years later during another melee on the subject involving the states and Congress:

Separations is, of course, far from being an exact science, and reasonable men will have honest differences as to the most appropriate formula for accomplishing a fair and objective cost allocation. . . .

At the same time, there is, in our opinion, little if any disagreement that there are certain minimum constraints of law and reason which must govern in any formulation of jurisdictional separations if it is to pass judicial review. This means, of course, that it is not possible to make arbitrary transfers of costs from one jurisdiction to another as a device to alleviate the need for rate adjustments. Instead there must be substantial and rational basis for the proposed transfer.

6

THE TARIFF SOLUTION

The first significant challenge to the Bell System's control of all aspects of the telephone business was, by anyone's description, a major one. In January 1949, the Justice Department filed a civil antitrust suit against AT&T and its huge manufacturing and supply subsidiary, the Western Electric Company. Selection of the U.S. District Court in Newark, N.J., as the location for the action reflected the suit's orientation to Western, since its largest plant was located in Kearny, in northern New Jersey.

Principal charge of the government against the two companies was establishment of a monopoly in the manufacture, distribution, and sale of telephone equipment through AT&T's near-100 percent ownership of Western. The court was asked to require divestiture of Western by AT&T, and the division of the manufacturing company into three regional firms, each independent not only of AT&T but of each other. Also asked was divestment by Western of its 50 percent stock interest in the Bell Telephone Laboratories (AT&T owned the other 50 percent).

The relationship of the operating telephone organizations, the Long Lines Department and the Bell System associated companies, to AT&T would not be changed. They would, however, have to buy their equipment through competitive bidding.

The government's complaint relied mainly on the FCC's findings in the special telephone investigation report. This was no coincidence. Attorney in charge of the case was Holmes Baldridge, then chief of the Antitrust Division's General Litigation Section, and former chief counsel of the FCC's telephone investigation in the 1930s.

Baldridge later made it clear, at congressional hearings long after he left the government and after the case ended with a consent decree, that the complaint had been largely his personal project. The only other attorneys involved, he said, were one who helped him on patent licensing matters and Assistant Attorney General Herbert A. Bergson, who approved the filing of the complaint and signed the document.

As Baldridge and Bergson envisaged the operation of competitive bidding after the breakup of Western, the Bell companies would issue specifications and data about their equipment needs. These would be aimed at enabling any qualified supplier to bid for sales on "a fair, non-discriminatory, and competitive basis." AT&T, Western, and Bell Labs would facilitate all this by licensing all patents and patent rights that they owned to all applicants.

The chief purpose of the suit, Attorney General Tom C. Clark said in a statement, was "to restore competition in the manufacture and sale of telephone equipment, now produced and sold almost exclusively by Western Electric at non-competitive prices."

Bergson, in a statement of his own, got to the crux of what would be a running battle among commentators on the later consent decree. He observed,

Although the Bell operating companies occupy the status of public utilities and hence are subject to regulation by state and federal commissions as to rates charged to subscribers, Western Electric is not subject directly to such regulation.

Telephone rates are fixed upon the basis of a fair return on the investment in the telephone plant, and where such telephone plant is purchased from a single concern, it is obvious that the prices for such equipment are not determined by competition in a free market. If the Bell operating companies are given an opportunity to purchase telephone equipment competitively, they will be able to buy such equipment at lower prices. This, in turn, would reduce both investment costs and operating expenses and afford state and federal regulatory commissions an opportunity to reduce telephone rates to subscribers.

Activist regulators had long tried to get a handle on Western's prices to the Bell companies, for the very reasons cited by Bergson. To assist the regulatory efforts of the state commissions, their national association, the NARUC, was in the process of setting up a special staff subcommittee to assemble and disseminate information and make reports on the subject. The FCC's special telephone investigators, as recounted in the first chapter, proposed to have Western made a public utility, but they had to settle for a watered-down cost-accounting legislative recommendation, which was ignored by Congress.

From time to time, a few states moved to disallow part of the Bell company's claimed rate base as representing excessive equipment charges by Western. But most states had settled into a fairly comfortable routine. AT&T had a price survey team at work keeping records of the prices that the smaller independent telephone suppliers charged non-Bell companies for the same or similar items. Western's prices were consistently lower. A member of the price survey team was a Bell witness in every state rate hearing to present the scorecard and to contend that telephone users benefited because of the Bell/Western relationship.

For the most part, regulatory commissions accepted the argument that telephone rates were lower because of the Bell companies' ability to buy equipment at lower prices. For the bulk of regulators, that took the issue out of rate proceedings. Only here and there were regulators heard to contend that they did not know whether, considering the size of its market, Western might, under tighter regulatory controls or general competition, be able to offer still lower prices. There was no argument about the quality of Western's products or what that quality contributed to the world's best telephone service.

There was no more deeply held article of faith by Bell System management than that continued vertical integration was crucial to good nation-wide telephone service. They believed that the Bell System existed only to provide good service and that Western was an essential means to that end. Rather than the operating companies being captive markets, they said, Western was a captive supplier. Their contention was that the telephone companies told Western what they wanted, and they got it.

Even when the regulators disagreed with Bell managers, they seldom seriously doubted the managers' sincerity. Total Bell control of the whole system was an accepted rule, which included the equipment that went into it. If the Bell companies could buy their equipment from Western at lower prices than they could get it elsewhere in the market that existed at the time, what was wrong with that?

For seven years, until the announcement in January 1956 that Justice and AT&T had agreed on a consent decree, the public heard nothing about the pending suit. There was no indication that Justice was going through the huge preparatory job necessary to bring such a gargantuan case to trial, either in the remaining four years of the Truman administration or after Dwight Eisenhower was elected president.

It turned out, in the later congressional hearings, that AT&T management had reason to eye Washington nervously. The company was served with "sweeping" interrogatories soon after the Korean War heated up in 1950, making it appear that the Justice Department was preparing for trial. As they began the massive job of complying, company officials felt they were facing one of the biggest antitrust trials in history. The 1981 antitrust case would later qualify as *the* biggest because of its wider scope.

When the Eisenhower administration took office in 1953, AT&T carefully noted the statement of the incoming attorney general, Herbert Brownell, Jr., that he was reviewing all pending antitrust cases. Discussions began, and AT&T urged dismissal of the complaint—for reasons including the national defense importance of AT&T to the United States, once again engaged in a war. Brownell made it clear that he was not about to dismiss the suit but said he was willing to talk about a consent decree.

The talks continued, off and on, from the spring of 1954 through the fall of 1955. It was not until December 1955 that Justice told AT&T it was ready to negotiate a decree that did not include divestiture of Western Electric. But for political cosmetics, Brownell urged that AT&T suggest some changes that could be made in Bell System practices that could form the guts of a so-called settlement.

A short time before, the FCC was engaged in correspondence with Justice which developed into a critical episode. Brownell wrote to the Commission saying that it had been alleged that divestiture would have a series of unhappy effects. Perhaps the most distressing, he had been told, was that if Western were no longer an affiliate of AT&T, the regulatory commissions would lose control of the prices paid for telephone equipment through disallowances from the telephone companies' rate bases.

Further, Brownell said, AT&T had pointed out that the prices charged by Western to the telephone companies were low, and that if Justice were able to split the manufacturer into three completely independent entities, equipment costs to the Bell companies would rise. This obviously would mean higher telephone rates. The FCC was asked to state its opinion about these consequences.

First cut at drafting a Commission reply was taken by the then-chief of the Common Carrier Bureau, Harold G. Cowgill. At a skull session, the FCC staff decided that Justice apparently was planning to dismiss the suit and wanted to use the FCC as a "patsy." Judging from the letter's style and content, the staff suspected that AT&T wrote the letter for Brownell. Cowgill's draft was cautious and inconclusive. It simply stated that if all those AT&T allegations were true, then in fact the effects of Western Electric divestiture on federal and state regulation would be adverse.

That, in the opinion of the commissioners, simply did not say enough about the powers of the regulators to control Western prices. FCC discussion of the subject was led by three commissioners who had previously served as state regulators and drew upon their experience in regulating telephone rates at both state and federal levels. All three were mentioned —one prominently—in subsequent hearings by the House Commerce "oversight" Subcommittee on alleged White House interference with the regulatory process, not the least of the political factors that later influenced the investigation of the telephone consent decree by a House Judiciary Subcommittee.

The Commission sent back the Cowgill draft for another try, and the job was assigned to Strassburg. He drew up a fairly comprehensive statement about regulators' powers to examine rate bases and the items that went into them, and to take whatever action seemed appropriate. Further, he pointed out, the state and federal regulators had used their authority to get a lot of information about Western's prices and, in using it, had been instrumental in obtaining some significant price reductions.

Two other paragraphs were included, in the interest of objectivity. One said the FCC was not in a position to know whether Western's prices were as low as they might be in a fully competitive market. With respect to the powers of regulatory commissions to deal with the problem, another noted, "It must be recognized, however, that the degree of effective application of these powers is largely dependent upon the resources of the respective regulatory agencies to examine and evaluate all the matters necessary to an informed determination of the reasonableness of Western's prices and profits."

Led by the three former state commissioners, the FCC struck those paragraphs. It was argued that the offending sections were "gratuitous and irrelevant," and that they related to how well the FCC did its job, rather than to its power to perform its functions. Thus, what went forward to Justice was a statement that the federal and state regulators were on top of the situation and fully empowered to control Western's prices under the regulatory setup as it stood.

Evidently, that was what Justice was looking for. By January 24, 1956, Justice and AT&T were ready to appear at the federal court in Newark with a consent decree and final judgment. The agreement limited the defendants to tariffed common carrier communications and work for the government. But it preserved the long-standing relationship among the manufacturing, research, and operating arms of the Bell System.

The key to the decree was that anything the Bell System companies did, except for government projects, had to be subject to regulation by federal or state authorities. If the function or service was covered by a tariff spelling it out, on file with public regulatory authorities, it was all right. It was this limitation that became a sore spot as the FCC opted in later years to deregulate what were traditionally treated as carrier services.

The tariff solution limited the Bell System to its then-accepted role, and the regulators were given a little assist in watching the levels of Western's prices. The decree provided the following: "Western is ordered and directed to maintain cost accounting methods that conform with such accounting principles as may be generally accepted and that afford a valid basis . . . for determining the cost to Western of equipment sold to AT&T and the Bell operating companies for use by them in furnishing common carrier communications services."

In its announcement, Justice made clear where it had been going. It said, "If Western Electric were to be divorced from the Bell System, the reasonableness of the payments made by the Bell System telephone companies for equipment and services would, in large measure, be removed as an issue in the ratemaking process. By leaving Western as the manufacturing arm of the Bell System, the regulatory bodies are left with adequate power to prevent excessive charges by Western."

Although Justice's 1949 petition to split Western from the Bell System had been abandoned, the patent licensing provisions intended to insure manufacturing competition survived in the decree. Bell's patent licensing policies had been fairly open, anyway, and the decree did not require much that was not already being done in making the products of Bell Labs research generally available.

The prize among Bell research for a host of manufacturers of electronic products was the transistor, probably the most revolutionary scientific development of its era. Until its invention at Bell Labs, television sets, radios, and a wide variety of other products had to be powered by vacuum tubes. The transistor virtually eliminated tubes, opening the way for much smaller and lighter products, easier and cheaper to manufacture and maintain.

The transistor was widely licensed by the Bell organization, and it opened the door to a whole new electronics industry. At the time of the decree, various small manufacturers told news reporters they were very happy with Bell patenting licensing as it stood. They said they were paying only about 1.5 percent of the selling price as patent royalties. Besides, they pointed to numerous meetings conducted by Bell Labs to give them the know-how to make transistors and said that was more important than the patents. All told, they credited the program with reducing the selling prices of transistors in four years by as much as vacuum tube costs had been trimmed after thirty years.

With the antitrust case settled and the decree a matter of record, the communications business returned pretty much to business as usual. Washington went back largely to politics as usual, with Democrats in control of Congress looking watchfully at the White House and waiting for the day when the first Republican President to be elected in twenty-four years would complete his second term. The next GOP candidate, they were confident, would not be a national hero like Dwight Eisenhower.

In matters related to communications, they found two targets. The first was Sherman Adams, former governor of New Hampshire, and powerful head of the White House staff as assistant to the President. Adams, who ran the day-to-day functioning of the executive branch of government, was not hesitant in calling regulatory commissions to ask about the status of pending matters. When he called on behalf of an applicant, the agency usually concluded that some action was immediately called for. In the government of those years, calls from or on behalf of Governor Adams were seldom ignored.

The House Interstate and Foreign Commerce Committee had the federal regulatory commissions under its legislative wing. Finding some indications of White House political favoritism—of particular interest to its Democratic majority when the incumbent President was a Republican—the committee set up a special investigating subcommittee with an aggressive staff.

It took a while, but the group finally put Adams on the road back to New Hampshire. The staff concentrated on his friendship with Bernard Goldfine, a financier with business before a number of government agencies. Adams had been a bit heavy-handed in dealing with matters, but the case was a long, long way from Teapot Dome or Watergate. Nonetheless, apparently Goldfine had once presented Adams with an expensive coat made of fur from an obscure South American animal. "Vicuna coat" became the symbol of the probe, and the outcry continued until Adams, still protesting his innocence, resigned.

The second and far less hard-nosed target was Richard A. Mack. "Richie" Mack was a relatively young, personable member of the Florida commission regulating public utilities, to which he was elected in statewide voting under the rules of that state. Naive, trusting, and out of his depth in Washington, he was moving up to become president of the NARUC, the state commissions' national association, when he was appointed to the FCC. The Eisenhower administration consistently found its federal regulatory appointees among the ranks of present and former state commissioners.

One of Mack's closest friends in Florida was Thurman Whiteside, a wheeler-dealer who became interested in an FCC proceeding in which the Commission was going to select a winner for TV channel 10 in Miami from among several fiercely contesting applicants. There was no evidence at all of outright bribery, but the Mack/Whiteside relationship had been such over the years that when Mack needed something—such as a loan to relieve a temporary financial problem—"Whitey" was there to help.

Totally unused to the national capital spotlight, Mack fell apart. His defense was weak: mainly that he and "Whitey" were "like brothers," but that this had no effect on his votes when Whiteside was involved in a contested case. By then, he was politically finished. Mack resigned and left Washington. Whiteside was prosecuted on several charges and eventually committed suicide in a jail cell. Mack fell out of sight for several years, returning to the news when he was found dead in a Florida transient room in an area described in some news reports as "skid row."

A few other members of the FCC figured briefly but inconclusively in the hearings. It was noted that former Chairman George C. McConnaughey had been an attorney for Bell companies in Ohio after his service on the state commission and before his appointment to the FCC. Some questions were leveled at his successor, John C. Doerfer of Wisconsin, centering on his alleged acceptance of hospitality and travel expenses from regulated entities. Nothing illegal was proven in either case, but in the atmosphere of the day the Commission's reputation in common carrier regulation became a trifle tarnished.

Investigating in its own area, the House Judiciary Antitrust Subcommittee staff did not fail to notice that McConnaughey, Doerfer, and Mack were the 1955 FCC members who had led the effort to tone down the FCC's reply to Brownell on the possible effect of Western Electric divestiture. For some time, the subcommittee had been probing antitrust consent decrees in the Eisenhower administration, and it held hearings in 1957 on an oil industry decree.

Now, Rep. Emanuel Celler, chairman of both the full Judiciary Committee and the antitrust unit, was ready to move on the telephone decree. "Manny" Celler was a skilled Brooklyn politician who was an institution in the House. He knew how to use the chairman's role to achieve the result he wanted, even if events would show later that, in this case, he did not have the support of a majority of either his subcommittee or the full committee.

The Celler hearings marked the first time in its almost quarter century of existence that the FCC's ability to regulate the Bell System was seriously questioned. Part of the Celler thrust was to put Justice Department officials in effect on trial for negotiating the consent decree. Where the FCC was concerned, it was to pit the Eisenhower administration appointees who were commissioners against the career members of the staff.

Celler's approach—and that meant the approach of the subcommittee staff who owed their jobs to him, however much other subcommittee members might protest—was two-pronged where the Commission was involved. First was the FCC letter assuring Brownell there was no real problem about controlling Western Electric prices as long as the existing Bell System structure was maintained. The other was the fact that, as recounted elsewhere, the FCC had always negotiated interstate telephone rates with AT&T and had never held a formal rate hearing.

When the Celler subcommittee staff began its inquiries at the Commission, Strassburg was assigned to provide the documents they asked for and cooperate otherwise. Chairman Doerfer was confident the Commission's record would stand scrutiny and that it had nothing to hide. Consequently, Strassburg felt "freer than I should have" to be entirely forthcoming in providing Herbert N. Maletz, chief counsel of the subcommittee, with candid answers to his questions. He did not know, among other things, that Maletz was recording the conversation.

When the subject turned to the letter assuring Brownell that regulatory powers were adequate to control Western Electric's prices, Strassburg described his original draft citing the practical limitations inherent in the regulatory process to deal with the problems of Western's prices and profits. He termed the FCC's final version of the letter as lacking "complete balance," and described the Commission letter as "something of a misrepresentation."

The Celler hearings ran off and on for nearly two months in the spring of 1958. At the beginning, Celler said the subcommittee was interested primarily in the way Justice discharged its responsibilities. Later, the scope grew much wider and took in the FCC.

In the first sessions, AT&T officials who had taken part in the decree negotiations defended both the process and the result. They firmly stated that divestiture of Western would "destroy" the Bell System as it was known. Among other things, they emphasized Bell's contributions to national defense over the years, and the fact that 8,500 patents held at the time of the decree were now being licensed, for all practical purposes, royalty free.

At one point, subcommittee counsel sought to show that substantively the decree and a list of questions raised by AT&T during the negotiations were remarkably similar. AT&T Vice President and General Counsel Horace P. Moulton pointed out that the crucial section of the decree restricting the Bell companies to regulated activities was not even hinted at in the AT&T document.

The sessions' political content was demonstrated regularly. Chairman Celler repeatedly ruled out order inquiries by the ranking Republican, Kenneth B. Keating of New York, questioning the validity of the original complaint. The subcommittee's purpose, Celler said, was to determine what led up to the decree, not the "merits of the case" as originally filed.

The FCC's appearance was an unusual one by congressional hearing standards. Not a commissioner was present. Strassburg, a middle-level official at the time, was the Commission's spokesman. He had been directly involved in the events being investigated, and presumably Celler and Maletz thought he would give answers they preferred. He was surrounded by higher-ranking staff members, including John Nordberg—who now headed the common carrier staff—but they were silent during the questioning.

Strassburg did his best to point out that just because the FCC had in its 1953 negotiations with AT&T (reported in chapter 5) acquiesced in rates producing 6.5 percent, it had not fixed that return as reasonable for all time. Celler, making arithmetical calculations, said that AT&T earnings of 7.3 percent to 7.8 percent in 1955 to 1957 showed that telephone users had paid $159 million more than they should have. Strassburg replied, "If you are going to take that 6.5 percent and freeze it and say that this is your point of reference for all time to come, you can sure come to that conclusion that rates are higher than need be by that amount. . . . If you want to carry your rationale forward, you might as well go back to the preceding years when [the actual rate of return] was less than 6.5 percent.''

There was little Strassburg could do about his recorded comments on the letter to Brownell except to reiterate his personal view that the final version lacked "objectivity" and "balance." But he reiterated that the regulators had considerable authority to disallow rate base items, and that many of them had used it. When Celler alleged that Western could charge the Bell companies "anything it wishes," Strassburg responded that the commissions "like to think the regulatory process keeps them from doing it."

All told, the hearings were probably a draw. When the subcommittee issued its report a year later, it was clear that no minds had been changed by the testimony.

The subcommittee had nine members, six Democrats and three Republicans. The vote on issuing the majority report was 4 to 3, with two Democrats not voting. Thus, the dissenters proclaimed, it was not being issued by a majority of the subcommittee.

Furthermore, there were indications that the full Judiciary Committee, although Celler was the chairman and the Democrats held a majority, might not have approved the document. The three dissenting members of the subcommittee reported that their "request to present the report to the full Judiciary Committee for consideration on the merits was refused. Therefore, the report has no status and is not a document of the House of Representatives."

Official or not, the report was sharply critical both of the decree and the FCC's regulation of AT&T. It was recommended that Justice move to reopen the case by asking the court in Newark, which retained jurisdiction over the decree, "for relief from the decree's inadequacies." It also called on the FCC to institute a "comprehensive, formal rate investigation to determine on a public record a fair return for telephone service."

The three Republicans charged that their colleagues essentially were just "second-guessing a lawsuit." While alleging bad faith on the part of Eisenhower administration officials, they said, the four Democrats were actually just disagreeing with their opinions.

Down the line, on every main point of contention, the subcommittee members disagreed. One warning of the Democrats did not stand the tests of hindsight and history: "By its enlistment of judicial confirmation, the decree constitutes a formidable roadblock to remedying monopoly in the affected industry."

But when they turned to FCC regulation of telephone rates, the Democrats were later to have the last word:

In sum, between June, 1955, and October, 1957, the Chief of the Common Carrier Bureau no less than six times called the attention of the Commission to the fact that the Bell System, by virtue of the 1953 rate increase, was deriving a return from interstate telephone service in excess of 6.5 percent. On at least two occasions in this period, the staff made the "very definitive recommendation" that action should be taken by the Commission looking toward a possible rate reduction. . . . It is significant that the Commission has neglected in the 24 years of existence to establish fundamental principles or standards by which to judge the reasonableness of the Bell System's interstate telephone rates.

7

CONGRESS IN ORBIT

Nothing had excited the imagination of the communications world more than the possibilities of the first earth satellites. It had long been formulated that when space satellites became a practicality, communications "relays in the sky" would be one of their primary uses. With experimental, early model satellites being demonstrated with some regularity in the later 1950s, communications planners moved to establish how the new medium would be developed and who would control it.

The glamor of the new technology and its potential for enhancing the communications networks of the world galvanized industry and government into action. In fact, satellite communications became the catalyst for one of the few times in the history of the Communications Act that the administration proposed and Congress enacted legislation to provide the national policy framework for institutionalizing technological change in the communications field. In doing so, they preempted moves of the FCC, which hoped to integrate satellite service into the established order without further policy direction from Congress or the White House.

The result was the Communications Satellite Act of 1962, with its creation of a chosen instrument—the Communications Satellite Corp., known as Comsat. The act was intended "to establish, in conjunction and in cooperation with other countries . . . a commercial communications satellite system as part of a global communications network, which will serve the communications needs of the United States and other countries, and which will contribute to world peace and understanding."

Political scientists to no less extent than pragmatic communicators have engaged in continuing, albeit academic, debate as to the extent to which these pious statutory objectives—particularly regarding world peace and understanding—have been promoted by congressional restructuring of the international communications industry.

Certainly, world peace and understanding seem no closer to fulfillment now than they did in 1962. In addition, there are legitimate grounds for questioning whether international communications are more or less efficient and economical with Comsat, whose principal role has been to lease to public service voice and record carriers access to communications satellite channels to use in conjunction with their terrestrial networks.

It did not take long to demonstrate that satellite communications, as a concept, was eminently practical. The first voice heard from space was that of President Eisenhower. A recorded message from "Ike" was broadcast during the 1958 Christmas season from an Atlas military satellite. Other demonstrations came from passive satellites, such as the giant Echo bag off which signals bounced as it orbited around the earth. Communications technicians, however, were looking primarily to active relays as the method to provide low-cost satellite communications service over long distances.

Everyone was getting into the act. In 1958, before any actual demonstrations, the President's Science Advisory Committee put out a report on the subject. It said satellites "could surely—and rather quickly—be pressed into service for expanding world communication." Transmitters, it was correctly forecast, would be powered by solar batteries which "should be able to keep working for many years."

Congressional committees, intrigued by the domestic and international political implications involved in any use of outer space, were looking for a piece of the new action. They were holding the beginnings of what probably was a new record for hearings by the most committees and subcommittees on a single subject. It sometimes seemed on Capitol Hill that, when two or more were gathered together, someone would call a hearing on satellite communications. One of the early starters was the House Committee on Science and Astronautics, which reported in May 1959 that the United States had reached the stage "where a useful worldwide communications system based upon the use of satellites can be initiated immediately."

Enthusiasm abounded. Later in 1959, the Army's Deputy Chief Signal Officer, Brig. Gen. Earle F. Cook, said in a speech to radio-television executives that the commercial possibility of a privately owned communications satellite system "staggers the imagination." He forecast that "although initially of considerable expense, the regular transmission of transcontinental video programs—and even transworld video programs—would soon absorb the cost of launching an active real-time communications satellite."

The National Aeronautics and Space Administration's (NASA) fast-moving space program included a major communications project. The military rushed to develop communications satellite plans and demonstrations.

The first active communications satellite, Courier 1-B, was a project of several Army Signal Corps contractors. It orbited at relatively low altitude. Courier, like all low-altitude satellites, flew in so-called random orbits, coming around the earth in varying routes. It was operable with U.S. earth stations only during portions of its orbits when it was in sight of the United States.

The alternative—then a well-established theory, and before too long a reality—was a synchronous satellite. Now in universal use, the synchronous satellite orbits at 22,300 miles above the equator, traveling at the same relative speed as the earth turns. Thus, it remains in the same position relative to the earth and is always available for communications among all of the ground stations visible to the satellite, hence, its decription as a geostationary satellite.

Communications companies, not wishing to be left behind in their own field, pressed research and development programs.

As the nation's predominant communications supplier, AT&T wasted no time testing the obviously tremendous possiblities of satellite service. In October 1960, two weeks after the first active communications satellite was launched, AT&T asked the Commission to approve satellite experiments it planned. (Transmitters aboard the satellite sent amplified received signals back to earth stations, as opposed to the bouncing effect of passive satellites.)

AT&T, naturally having primary interest in satellites for telephone service, was an early and vigorous advocate of random satellites and opponent of the geostationary approach. As to the latter, it kept pointing to the six-tenths of a second delay in the time it would take a signal to make the nearly 44,500-mile round trip to the satellite and back to the receiving ground station.

Delay caused by the lengthy transmission path, AT&T emphasized, would result in an echo problem despite the best echo control devices then available. If an extremely long international call required a second satellite relay to cover the distance, the echo delay problems would double, AT&T engineers pointed out.

Compared to AT&T's random orbit preference, synchronous satellites had obvious advantages in efficiency and cost of operations as a whole. The argument was therefore a relatively short-lived one. From its establishment, Comsat opted for synchronous satellites. The delay problem has proved to be at least manageable and tolerable in voice service and, of course, has no relevance in some other forms of communications for which satellites are heavily used, such as television.

Even more contentious was the struggle over who would own and control the international satellite system. It was virtually taken for granted that the new technology would have its most immediate and beneficial application in commercial operation of a worldwide system of satellite communications. Here is where, it was felt, the United States could most effectively use its technological leadership in satellites to garner global goodwill by providing a much-needed improvement in quality, availability, and economy of international communications.

The industrialized nations of the world, by and large, had enjoyed quality communications with each other via modern terrestrial cable and radio facilities. Satellites could extend the same advantages to the underdeveloped nations, who were mainly dependent on relatively inefficient and unreliable high frequency radio for their voice and record communications with the United States and other countries.

No less intriguing was the prospect that satellites offered for the transmission of real-time television, sending programs and news events over vast expanses of land and water—something impossible to do with the then-existing communications networks.

In keeping with tradition, U.S. industry officials firmly believed that private enterprise should be responsible for the commercial development of the new technology. On the other hand, there were many Americans who regarded communications satellites as not just another commercial innovation. They viewed satellites as the product of a space program and related research financed by the general taxpayer, who should be the one to reap the dividends from these advances. Moreover, they observed, satellite communications were uniquely suited to U.S. foreign policy objectives of advancing the peaceful uses of outer space.

Thus, there were a number of influential and vocal advocates of government maintaining a strong grip on the use to be made of this national resource. The intensity of the debate over how to structure the government's relationship to the technology was, of course, heightened by the assumptions that economically and operationally there was room for only a single global satellite system.

Not surprisingly, the FCC was quickly found in the vanguard of those taking the traditional free enterprise approach. At the same time, the Commission advocated that the new technology be institutionalized with the minimum structural impact on the existing international communications common carrier industry.

The Commission viewed satellites as just another, however advanced, means of providing public communications service. In this, it was buttressed by an agreement entered into in 1959 with the space agency, NASA, defining the agencies' respective roles and responsibilities in supporting satellite communications experiments. In effect, the accord provided that NASA would be responsible for launching satellites, and the FCC would license and regulate the commercial communications interests that would be involved.

In March 1961, the FCC embarked on a formal public inquiry into the ownership and operation of a global satellite communications system. Two months later, it arrived at the tentative conclusion that satellite technology could and should be integrated into the existing international communications networks of AT&T and the international record carriers, as the ones best qualified to do the job. In the Commission's perception, this could be achieved without the need for any new policy direction from Congress, and within the existing framework of the FCC's licensing and regulatory powers under the Communications Act.

Soon thereafter, President John F. Kennedy took away some of the FCC's intitiative. He issued a policy statement of his own, declaring that the national policy was one favoring private ownership and operation of a communications satellite system under a set of specified policy requirements.

The following day, the FCC called upon the international carriers to promptly get on with the business of organizing themselves into a consortium for the joint planning, ownership, and operation of the satellite system. Arrangements were to be made for ownership participation and nondiscriminatory access by foreign communications entities. In short, the FCC contemplated that communications satellites would be integrated into the established international global network, following the same conventions and practices that for generations had governed the ownership and operation of international cable and radio networks.

At issue was what was developing into a fundamental national debate. At one extreme was the established communications order, spearheaded by the FCC and the existing common carriers. To most, the existing common carriers really meant AT&T. They accepted and espoused the concept of a regulated natural monopoly, controlled by the government in the public interest. Willing to concede the potential dangers of most monopolies, they felt that the traditional warnings about them just did not apply to communications.

They pointed to the quality of service available to U.S. citizens, at rates which had decreased or at least were more stable than consumer prices in general, as evidence that the monopoly bugaboo did not apply to communications. Besides, the existing system was the one they were familiar with, and the one they had made work.

At the far opposite end was an increasingly vocal minority who simply could not buy the precept that any monopoly was good. To them, all monopolies were dangerous and at the very least needed strict government oversight. They did not believe government control of the communications carriers, especially AT&T, was tight enough, and they certainly did not wish to extend the AT&T monopoly into outer space.

Besides, they reiterated frequently, the whole space program had been carried out by the government and paid for by the taxpayers. Benefits of commercial application, especially communications, should go to the taxpayers, too. Those monopoly communications common carriers—again read AT&T—should not be allowed to come along now and pick off the fruits. They urged government ownership of the satellite communications system.

In between were the pragmatists of the Kennedy administration, whose interest lay in making U.S. participation in satellite communications work, not in making philosophical statements for or against the existing system of regulated monopoly. At the same time, the administration could not ignore the political implications presented by AT&T's dominance in the communications industry and the regulatory shortcomings of the FCC. Therefore, the administration bill sent to Congress in early February 1962 was a blend of the disparate views raised in and out of Congress over the preceding several years.

The White House proposed a satellite organization owned and operated by private enterprise, but under strict government supervision, including some at the Presidential level. Among other things, several directors would be Presidential appointees, confirmed by the Senate. Government agencies in the space and communications areas, notably NASA and the FCC, would retain their existing authority, with some augmentation.

The space communications company would have two classes of stock. Class A would be open to subscription by anyone and would have the traditional rights of common stock. Class B stock could be acquired only by common carriers. It would not be entitled to regular dividends and would have no voting rights. The carriers could, however, include their stock investment in their rate bases to the extent allowed by the FCC. It was hoped this would induce AT&T and the record carriers to lease satellite circuits from the new carrier's carrier.

There followed an extraordinary congressional debate which lasted for more than six months. Despite the extensive hearings already held, before final enactment of the bill, five different Senate committees and subcommittees, as well as the House Commerce Committee, conducted further hearings on the legislation. They were often long and, at times, loud.

During much of the debate, a bystander might have thought that the proposed legislation provided for AT&T as the government's "chosen instrument" in space communications, rather than the organization that became Comsat. Primary objection of the dissenters centered on what they viewed as the likely control of the whole operation by AT&T as well as its enrichment by publicly financed space experiments.

When the hearings were held, the FCC did little for its declining popularity by continuing to push for legislation which would give the established carriers—again, this meant AT&T to the opponents—a greater role in satellite communications. The concept of Comsat-to-be in the administration bill was that of a "carrier's carrier," providing the U.S. portion of space segments which the carriers could use in furnishing their authorized services. Throughout the debate, the administration approach was accepted by a majority of the legislators.

Other members of Congress were dropping their own versions of space communications legislation into the hoppers, particularly in the Senate, which developed into the main battleground. One proposal was that of Oklahoma's Robert Kerr, chairman of the Senate Aeronautical and Space Sciences Committee.

Senator Kerr did not like the financing provisions of the adminstration bill. His version would give the common carriers buying stock more of the traditional rights of stockholders. Under the administration bill, he argued, the carriers were being invited to put up half of the initial capital of the space corporation, but to get little or nothing in return for their money. It was his version which came closest to the final law.

Senator Kerr's committee held the first set of hearings on the administration bill and produced the compromise legislation which finally passed. It varied from the measure that had been sent up from the White House mostly in the financing provisions. Among other things, instead of being rigidly segregated into two classes, the stock would be issued in two series: series I for general investors and series II for common carriers.

Meanwhile, the opponents were busy. A Senate bill providing for government ownership of any U.S. portion of a satellite communications system was introduced by two Democratic senators, Estes Kefauver of Tennessee and Wayne Morse of Oregon.

Throughout the six-month battle, traditional political lines were blurred. The bill backed by the Democratic administration of President Kennedy, with the Kerr financing changes, received the solid and essentially unanimous support of Republicans. The die-hard Senate opponents were all Democrats.

Most of the opponents were in the group normally described as "liberals." One of their most active recruits, however, was usually a strong conservative in business and economic matters, Senator Russell B. Long of Louisiana. Perhaps reflecting his populist family background—his father, Sentor Huey Long, had controlled Louisiana under the slogan "Every man a king"—he was a stalwart among the dozen or so senators who battled against the administration bill. It was Senator Long who gave the dissenters their rallying cry, "If the FCC cannot control AT&T on land, how in the world can it do it in outer space?"

On the House side of the Capitol, the administration was in control. The House Commerce Committee held relatively brief hearings and then seemingly waited for the Kerr compromise to come out of the Senate Space unit.

Several weeks after it did, the House committee favorably reported a bill closely similar to the compromise.

The administration, meanwhile, had seen how the legislative dynamics were working. It had dropped its opposition to the Kerr financing changes in the compromise bill and officially got behind the revised version.

The House presented no problem to administration backers anxious to demonstrate another national commitment to the space program. The House bill, little changed from the one reported by the Interstate and Foreign Commerce Committee, rolled through by a 354 to 9 vote after two days of floor discussion. During the debate, a proposed government ownership amendment was rejected by an overwhelming vote.

The Senate was a different story. There, the battle was in the hands of legislative tacticians.

What faced the Senate majority leadership was the omnipresent threat of filibuster. Here again, the satellite legislation produced some strange alliances. At the time, the nation was still in the throes of civil rights struggles. The liberals opposing the bill normally were strongly against the use of the "extended debate," or filibuster, tactic, by which a determined minority could bring the Senate to a halt and force withdrawal of a bill to which they were unalterably opposed. In the satellite communications instance, they were threatening to embrace filibusters as though they had created them in the first place.

In 1917, the Senate had adopted the "cloture" rule, under which the Senate's rules permitting unlimited debate could be suspended. Cloture had been invoked successfully only four times since adoption of the rule, and not for thirty-five years. A move to invoke cloture always brought about some strange voting patterns and clearly would in this instance. Southerners who were supporting the satellite bill would rather be caught dead than voting for cloture. On that vote, the bill's opponents would pick up significant support.

Since cloture at that time required a two-thirds vote for approval, the tactical weapon in the hands of the opponents was considerable. Congress was looking to adjournment in time for the fall election campaigning. If enough delays could be introduced, the satellite bill would be passed over until the next year, and heaven knows what might happen then. The effort almost succeeded.

In late May, the crucial Senate Commerce Committee, in addition to Space, favorably reported the bill supported by the administration. Two of its members objected, and the delaying tactics were in full bloom. They obtained a ten-day delay in reporting the bill to give them time to write minority views.

In mid-June, the Senate first took up the bill. Rather, since legislative tactics were paramount, it took up a motion to make the satellite act its pending business. The motion was subject to full debate.

Senators Kefauver and Long led the charge. Among other things, Senator Long asked, "Who besides AT&T wants this bill?" Senator John O. Pastore of Rhode Island, floor manager for the bill, replied, "The President does." Over the next week, the "extended debate" continued, and the Senate was in a parliamentary vacuum. Lengthy speeches were made by the opponents who insisted they were not conducting a filibuster. Again, it was not the bill itself which was being debated, but only the motion to take it up for consideration. For a time, whatever it was called, it was successful.

The opponents offered thirty-seven amendments, each technically subject under Senate rules to unlimited debate. Senator Kefauver offered his bill, providing for government ownership of the U.S. space segment, as a substitute, with seven cosponsors. Senator Long argued that the FCC had shown itself unable to regulate common carrier rates and thus should not be entrusted to such a role where the space system was concerned.

In the process, seven members of the dissenting group proposed a new FCC telephone investigation, with an appropriation of $3,265,900. It called for a report to Congress in a year and a half of all of the events since 1939 relating to AT&T rates.

Senator Kefauver's Judiciary Antitrust Subcommittee had held hearings earlier in the year, ostensibly on the satellite bill but mostly focused on FCC regulation of AT&T rates. Now the subcommittee issued an appendix to the printed record of its sessions. It included a variety of internal and heretofore unpublicized FCC records, showing that the Commission had, during the preceding year, turned down two proposals for formal, public AT&T rate investigations.

In July 1962, the Bell System gave the fledgling satellite communications program its biggest technical and public relations boost. To widespread acclaim and enthusiasm, the Telstar experimental satellite was a complete success. Moving through space at 15,000 miles an hour in 160-minute orbits some 7,000 miles above the earth, it carried live transatlantic TV pictures and "bread and butter" services like telephone, data, and facsimile whenever it was visible to the first earth stations. They included stations built by the Bell System at Andover, Maine; one in France; and another in Great Britain.

A global system of this mode would require the launching of forty to fifty satellites and would involve two earth stations in each country in order to acquire and track a given satellite for continuous communication. The complexity and cost of this concept was ultimately found unacceptable and had to yield to the geostationary approach.

The country was still in the midst of Telstar mania when a motion to make the satellite bill the pending business of the Senate was again called up for discussion in late July, two weeks after the Telstar launch. It became evident quickly that Telstar had not changed the political situation. The opponents' first move was to demand that the journal of the previous day's Senate proceedings be read in full, something almost always dispensed with by unanimous consent. Senator Morse, holding the floor at the time, even threatened to offer amendments to the journal.

During the next week, a short-term compromise was reached. The Senate agreed to a one-week referral of the legislation for hearings by yet another committee, Foreign Relations. The opponents, arguing that Foreign Relations' expertise was needed to consider those aspects of the projected international consortium, had urged such a move.

After a week of sessions, including an unusual one at night, Foreign Relations favorably reported the bill back to the Senate by a 13 to 4 margin. It had rejected thirteen proposed amendments in the process.

> *Authors' comment:* During this time, we had another lesson in the workings of Washington. Innumerable questions by Senator Long had slowed the hearing to the point where the night session was ordered. We had dinner at a nearby restaurant. Gloomily anticipating another session of indeterminate length, we noted Senator Long dining with another man at a nearby table. Leaving the restaurant, we asked the Senator if he were going back to the hearing. "Oh, no," he replied cheerfully. "My friend and I are going to the ball game."

By August 10, the Senate was in its fifteenth day of debate on the bill, after the hearings by five Senate committees. Senate Majority Leader Mike Mansfield warned that the rare step of cloture would be sought if a vote were not permitted. Early the following week, the "extended debate" was continuing, and the leadership put the cloture procedure into effect. The Senate was still being told that the bill would "ensure AT&T's continued monopoly."

The cloture vote was the dissenters' high-water mark. Joined by die-hard opponents of closing off debate in the Senate, mostly Southerners, they lost by a 63 to 27 tally, four short of what they needed. From then, it was all downhill, and the Communications Satellite Act of 1962 passed the Senate by a 66 to 11 vote. The bill was little changed from the one adopted by the House, which bypassed the necessity of a joint House-Senate conference to iron out the differences. The House simply adopted the Senate-passed bill by 372 to 10.

Only one of 230 amendments offered in the Senate was adopted. It seemed fairly routine at the time but would become more significant in the middle 1980s when the competitive era spawned proposals for additional satellite systems. Offered by Senator Frank Church of Idaho, the amendment was aimed at insuring that nothing in the bill would preclude establishment of an alternative satellite system in the future, if such action was considered to be in the national interest.

After signing the bill into law, President Kennedy nominated thirteen incorporators to set up the new satellite communications company. One of their early actions was to decide to call the company the Space Communications Corp. It developed subsequently that some far-sighted operator had registered that corporate name in the District of Columbia, where the company was to be headquartered. Evidently he wanted too much for the rights to the name of the company. It quickly became the Communications Satellite Corp., or Comsat.

Perhaps never has there been such a rush to buy the stock of a company which had no assets except a charter. Some members of the public were convinced that Comsat stock was the road to riches. Having listened to all the optimistic forecasts, some reporters who had covered the debate—generally not sophisticated investors—announced they would buy a few shares and send their children or grandchildren through college on them.

Comsat also looked good to the common carriers eligible to buy half of the authorized 10 million shares. The FCC set up a procedure for eligibility, announcing that any common carrier subject to its jurisdiction, including small telephone companies, would fill the bill. The Commission received 210 applications, and when the stock was offered in mid-1964, 163 applicants were in line with checks in hand.

The 5 million shares of series II (common carrier) stock were oversubscribed by one-third, and an allocation formula had to be used. AT&T, noting that it used 85 percent of total international communications capacity, asked permission to buy 85 percent of the stock, or 4,250,000 shares. The allocation formula was applied mostly to it, and it wound up with 2,895,750 shares. Others subscribing to 5 percent or more of the available stock were the International Telephone & Telegraph Corp., General Telephone & Electronics Corp., and RCA Communications.

There were rumors that the White House, now in the hands of Lyndon Johnson, who as Vice President had headed the U.S. Space Council, had made clear its interest in full subscription to Comsat stock. The reports were that the President was concerned that if the big communications companies did not buy up their full allotments, individual investors would be worried when the 5 million shares for the public hit the market.

If so, the fears were groundless. The public quickly bought up its stock at the same $20-a-share price, and many members of the underwriting syndicate had to use an allocation formula because public demand was so heavy. Some brokers reported they would deliver one share for each ten their customers wanted to buy. The average purchase per customer was reported to be ten shares, as a result of the allocation. On its first day on the over-the-counter market, the stock, sold at $20 a share, hit $28. It settled down later to about $23.

Many of the fears and expectations which surrounded Comsat's beginnings never materialized. AT&T's stock holdings allowed it to nominate only three of the fifteen members of the Comsat board of directors, and there never were any visible or even rumored signs that AT&T was dominating Comsat.

On the contrary, because Comsat's top management was almost totally inexperienced in the practicalities of building and operating a communications system—particularly one requiring interfacing with domestic and international terrestrial networks—AT&T's guidance and advice were of tremendous value to the neophyte corporation.

From the outset, there were also constant suspicions that, notwithstanding their stock ownership in Comsat, AT&T and the international record carriers would give preference to the use of the submarine cables which they owned and minimize their use of satellite channels which they could only lease. The carriers, it was noted, could earn a return on their investment in their cables as part of their rate bases. Rental payments to Comsat, on the other hand, could only be recovered dollar for dollar and were no source of profit to the carriers.

To assuage those concerns, the FCC exercised tight controls over the building of new international cables and took additional measures to promote reasonable parity in the numbers of cable and satellite circuits to be activated by each international carrier.

All the warnings about AT&T domination of Comsat had not been lost as a political phenomenon, however. The frequent congressional complaints about the absence of an FCC rate investigation were to be answered in a few years. On the international scene, the Commission demonstrated in a little less time that it had received the message.

AT&T and its overseas partners in telephone service had filed with the FCC a proposal for the newest and most modern transatlantic submarine cable, known as TAT-4—ownership of which would be shared by AT&T and the west European government communications administrations. The cable would add significantly to transatlantic capacity, and its voice circuits were equally useful for record (telegraph and telex) and data services. In keeping with past practice, AT&T was prepared to provide the U.S. record carriers with long-term leasing arrangements in the cable.

The competing international record carriers—RCA Global Communications, ITT World Communications, and Western Union International—were joined in strongly expressed fears by the FCC staff. The concerns were that an AT&T cable providing circuits for alternative voice/record-data service would increase AT&T's share of the record/data business and endanger the viability of the international record companies.

The FCC readily accepted the recommendations of its staff, which were designed to curtail extension of AT&T's dominance of the domestic market into the international arena. They were that the TAT-4 cable approval should be conditioned on a new policy. AT&T would be left firmly in the international voice business but would not be allowed into the new record/data services. Those would be the preserve of the international record carriers. As an added measure to curb AT&T's international presence, the company had to share ownership to the U.S. half of the cable with the record carriers.

Loudly voiced fears of the satellite bill's opponents that common carrier stock ownership in Comsat would lead to the carriers' domination of the company were among the alarms that did not pan out. Comsat stock moved up nicely on the market, and the carriers had other uses for their money. Between 1968 and 1970, the stock sellers included ITT, GTE, and RCA, and the carrier proportion of total Comsat stock dropped from the original 50 percent to 30.8 percent.

AT&T, as always, was in the peculiar position that came from its sheer size. It owned 29 percent of all Comsat stock, as well as 99 percent of that left in the hands of common carriers. For it to "dump" its stock might be catastrophic for Comsat's financial viability, at a time when the global system was still being developed.

Besides, as a substantial user of Comsat's overseas services, from which it derived an increasingly large number of its circuits, AT&T had valid reasons for wanting Comsat to succeed as the nation's "chosen instrument" in the satellite field.

By 1972, as will be covered in chapter 12, the FCC was considering applications for domestic U.S. satellite systems. AT&T and Comsat had each applied for authority to set up domestic systems. AT&T, however, planned to take advantage of the experience Comsat had gained by leasing a satellite system segment from Comsat for its authorized domestic services. Later, the two companies would also be competing for specialized or private line services. The FCC concluded that AT&T's ownership of a significant chunk of Comsat stock would not, under all the circumstances, be conducive to a healthy market environment.

The Commission approved the lease arrangement but only on condition that AT&T divest itself of its ownership of Comsat stock. Meanwhile, the FCC, for the usual pro-competitive reasons, put other restrictions on AT&T's domestic satellite authorization. The most significant was that AT&T had to give its competitors, including Comsat, a three-year head start in providing private line and other specialized services.

AT&T needed no prodding to sell its Comsat stock at a profit exceeding $100 million, more than double its original investment. In doing so, it pointed out that conditions had changed greatly since 1964. "Since that time," it said, "Comsat has developed its own expertise and is a viable entity in its own right, thus obviating the need for the internal guidance and assistance of AT&T and other carriers."

Comsat has succeeded modestly as a going business, but it did not turn out to be the gold mine envisaged by early stock buyers. Its earnings were regulated by FCC in the traditional manner, and after a long and complex rate proceeding, it had to refund what were found to be excess earnings by the Commission. Its customers, the carriers using its services, in turn had to "flow through" the refunds to their customers. In 1987, another big refund order was issued by the FCC.

The average stock buyer in 1964 put up $200 for ten shares. There was a two-for-one split, so the buyer then had twenty shares, worth about $550 on the market in 1987. Those twenty shares paid a modest dividend of $24 a year, at the time of a proposed merger with Contel Corp., later terminated by Contel. Respectable enough for a $200 investment, but a child or grandchild going to college in 1987 would need some other sources of financial support.

With the passage of time and, particularly, as competition and deregulation have emerged more and more as the touchstones of policymaking, there is increasing sentiment to strip Comsat of its protected status as a statutory monopoly.

8

THE BIGGER THEY COME . . .

As they do in many lines of business, large customers present the biggest opportunity, and the biggest set of problems, to communications organizations. Obviously, the big dollars are there, and when they buy service at the same rates as small users, the large customers provide bigger net profit margins, as the result of economies of scale. At the same time, their very size and purchasing power put them in a position to demand volume rates and to at least threaten, if the numbers prove out, to "do it themselves" by building their own systems.

The Communications Act's prohibition of unreasonably discriminatory rates has consistently given regulators their biggest headache in dealing with proposals of the communications carriers said to be necessary as a way to meet competition. The struggle at the FCC has gone on full-scale since the middle 1950s over a variety of tariff offerings, usually by AT&T, aimed at attracting the big customers and keeping them from putting up their own communications networks.

The heated and never-ending disputes have centered on the proper way to allocate AT&T's costs among its services, particularly those used by residential and small business subscribers and those that meet the needs of big, or at least bigger, business. The battleground has also included proposals by users—mostly large, but often medium-sized if they band together in shared systems—for authorization to use the public's radio spectrum for their private systems.

As long as there is any government regulation or control of the communications business, the battle is sure to go on. The arrival of competitve long distance carriers on the scene greatly widened its scope. They, too, must be regarded as big customers of the established or dominant carriers by reselling AT&T services, using its long distance system to reach the many less-profitable areas of the country, and necessarily relying on the large exchange telephone companies as essential links in the chain of service.

For the large masses of residential and small business users, the local exchange carriers will likely retain their role as the "only telephone company in town." But as in the past, it is the large volume users who have the alternative means of service. Either singly or in such aggregations as shared tenant systems in building complexes, they are the ones who can bypass the local carriers, putting satellite antennas on building roofs, becoming customers of large "teleports," constructing their own microwave systems, or leasing private line services from AT&T and, increasingly, its competitors.

In the history of communications regulation, the FCC's private microwave case was a major watershed. It sounded the final notes for a long and deeply held view that if the regulated monopoly provided adequate service at reasonable rates, there was no compelling reason to allow other entrants into the market.

The case had its foundations in all that wartime research into broadening the useable boundaries to the radio spectrum, and in the Bell System's postwar demonstration that microwave was an effective, reliable mechanism for commercial-grade service. For private users, design flexibility was another advantage, since microwave systems can be engineered for anything from a modest to a very heavy traffic load.

Construction cost, compared with digging a cable or putting up a pole line, was equally important. In flat terrain, microwave towers on building tops or heights can "look" at each other at a distance of more than forty miles, and the owner or operator need not worry about owning any of the land in between or acquiring rights-of-way. All the owner needs is an FCC license to use the appropriate radio frequencies.

In November 1956, the Commission decided that what had been a piecemeal approach to possible private use of microwaves, the frequencies above 890 megaHertz—were known at the time, more comfortably and conveniently, as megacycles—should take the form of an overall study of the subject. The following year, fairly extensive hearings were held. The issues included several involving competition between private systems and common carriers. The carriers argued that private microwave systems would be an inefficient use of the radio spectrum—a scarce national resource—and that use of those frequencies solely by common carriers would serve to benefit everyone. In fact, AT&T had been integrating microwave technology into its intercity networks since the end of World War II.

On the economic side, the carriers contended that the great bulk of users of their systems would suffer if the big customers could go off on their own. Therefore, they urged that private systems be restricted to right-of-way companies such as railroads, which already had private wire systems along their lines, or to isolated areas.

The contention of economic effects is one which remains familiar today and no doubt will last as long as the industry. If big customers are encouraged to leave the common system, which is intended to serve all users, its support will be left increasingly to smaller entities who cannot afford to provide their own systems. Rates will have to go up as volume declines, sending more medium-sized customers, who had been on the fence about a move, off the network. Eventually, only individuals and small businesses will be left, scrambling to pay still-higher charges for deteriorating grades of service. Today, this is known as bypass.

During the mid and late 1950s, individual companies, national associations, and microwave equipment suppliers, whose interests favored consideration of private systems, had been pressing the FCC to authorize them. The lead cases were applications by Central Freight Lines of Texas, which wanted to dispatch and monitor its trucks between Dallas and Fort Worth via a private system, and the Minute Maid Corp., planning an extensive private system among its citrus-growing and -processing properties in Florida.

While the overall inquiry was going on, separate hearing proceedings were held on the two pilot applications. Issues were the same as they were in the omnibus private microwave case, and it became increasingly evident that the answer lay in an overall policy decision.

The internal struggle at the Commission was intense. A basic philosophy, which had held sway in utility regulation since the dawn of the century, and at the FCC for more than twenty years, was being challenged and in the process of being swept away. When, in August 1959, the FCC issued its final report and order in the "above 890" (megacycles) case, the Central Freight and Minute Maid applications had been pending for half a decade, the overall inquiry was nearly three years old, and it had been close to two years since the hearings were held.

Theoretically, just about anyone was now eligible to build a private microwave system. The FCC provided that those eligible for its safety and special radio services—the mobile systems familiarly used in taxicabs, police and fire vehicles, delivery services, and a vast variety of others—could obtain licenses in much of the nongovernment frequency space above 890 MHz for private point-to-point, "operational fixed" microwave radio outlets.

Since the safety and special services included such open-ended activities as business and citizens' band radio, that meant, in theory at least, that about any U.S. citizen without an extensive criminal record could, if he or she wished and could afford it, build a private system.

The common carriers lost on all of the main points they had made. Adequate frequency space was available, the Commission ruled. It made no difference, it stated, whether adequate common carrier services were readily available or not. And the FCC made clear it would not even consider, in acting on applications for private systems, what the economic impact on the carriers and their customers would be.

Not to worry, though, it told the common carriers, their customers, and the state regulators who were taking an increasingly dim view of the FCC's tendency to chip away at the telephone companies' business. (The state commissioners basically were concerned more about their own situations than about those of the telephone companies. They held office to regulate utilities, including telephone companies, but not private systems. The best way to continue to hold office was to seldom, if ever, raise basic rates for residential service, except as a last resort. A telephone company with a hold on most of the business was less likely to propose a general rate increase, and if it did and proved its case, most of the needed revenue could come from sources other than Aunt Minnie.)

The common carriers would not lose much business, the FCC said, because the cost of private systems was too large to attract more than a few system builders. Besides, the carriers had regulations in their tariffs limiting interconnection of private systems which would act as deterrents to the private networks.

In addition, the Commission pointed out, its decision was not throwing the doors open wide to all sorts of cost-sharing cooperatives which could be expected to greatly increase the use of private systems. Only right-of-way companies and those whose rates were subject to public regulation, such as railroads, power and pipeline companies, airlines, and truckers, would be allowed to band together into cost-sharing groups.

It was true, of course, that the nation did not immediately sprout private microwave towers on every hill and on top of every high-rise apartment and office building. Communications systems cost money, and they are not simple to operate. The decision to build an extensive private system is a major one, and a manufacturer or a food processor does not lightly enter the communications field. Communications carriers can be expected to furnish the service more conveniently and possibly at lower cost, if they can design a tariff offering that will stand the scrutiny of the regulators.

But even without vast arrays of private microwave systems, the significance of the decision was manyfold. The traditional regulatory premise that all large customers must be kept on the common user network for the public good of all users was substantially diluted. Of course, they would remain users of common carrier services if it was to their economic benefit. But they now had choices.

The choices were not all-or-nothing alternatives. The Commission's 1959 assurances that the economic danger to common carriers was minor quickly disappeared. The carriers could not sustain their tariff restrictions on interconnection for long against the pressures that mounted. The Carterfone decision in 1968 gave big users the right to interconnect private systems to AT&T lines, extending their private networks to places where they chose not to build. If a company had installations around the country, it did not have to put up a nationwide private system—links between a few key cities might satisfy its purposes.

The limitations on sharing of common carrier services and private systems also were steadily chipped away. Interim restrictions on private cost-sharing cooperatives vanished as innovative managers and promoters found ways around them, and any effort to restrict sharing and resale of services finally collapsed in general permissiveness. In the old days, when rate levels and relationships were established more on the basis of value of service than on cost, resale and sharing had been strictly verboten by the telephone companies. In the bright new world of competitive economic theory, every service is to seek its own cost level, regardless of the way it is used, shared, or resold by the customers of record.

But perhaps the greatest importance of the private microwave "open door" was the club it put in the hands of the big customers. It forced the common carriers—in the early days mostly AT&T and to a small extent Western Union, and later the myriad competing "OCCs" (other common carriers)—to come up with service and pricing plans to meet competitive necessity.

The club was to some degree effective as well against the FCC. Urged on by the congressional mandate against unreasonably discriminatory or preferential rates, as well as AT&T competitors who found discrimination in every AT&T move to keep its big customers, the agency found the established carriers' pricing responses at best questionable, and often concluded they were illegal.

Simply put, the club was, "Approve this tariff which gives us the lower rates we deserve, or we'll build an extensive private system and a large part of our communications expenses will no longer be part of the common carrier 'pot.' " It was used in FCC hearings by the Commission's sister government agencies, collectively the largest user of AT&T's private line services (those other than the familiar, "regular" telephone call).

Next largest group of private line users was composed of the nation's airlines. Their communications arm, for planning, engineering, and analysis of all services, and the operation of the airlines' privately licensed radio facilities, is Aeronautical Radio, Inc., or ARINC.

ARINC is essentially a nonprofit association of the nation's airlines, to serve as their communications manager. Established in the early years of the FCC and in the relatively early period of widespread air travel, it was founded on the basis that airline safety and efficiency would be enhanced if a professional organization took over the job of running the communications links necessary to control and monitor aircraft operations at airports and in the increasingly crowded air space.

When ARINC was founded in the 1930s, it was an exception to the general prohibition against sharing and joint use. But its founders convinced the FCC that the then-pioneering air travel industry was an exceptional case and that something special was required. Consider the chaos that might have erupted if a separate communications group from each airline had been fighting for space in the radio airwaves and in the airports then just being built. From a communications standpoint, the airline industry became unique and stayed that way.

In the turbulent years for large users which followed the private microwave decision, ARINC, to do its job, had to have a private microwave system at least under study. Tariffs which promised big customers the opportunity to make maximum cost-effective use of their communications services had to be constantly weighed and analyzed against the now-available alternative. Cost and feasibility of large regional and even national private networks, possible network design, availability of equipment, and numerous other questions were studied by ARINC committees for years.

As a result, ARINC could tell the FCC repeatedly, in the single-customer role which the airlines enjoyed alone among U.S. private industries, that any regulatory decisions ruling out bulk service offerings or requiring the same pricing techniques as small user offerings would inevitably lead to a private system.

One bulk pricing plan stood in the vortex of the controversy over private line ratemaking which raged around it for nearly twenty years. It galvanized the use of communications among large U.S. companies. It was a benign giant to its customers, promising and delivering all sorts of new benefits. It was a huge menace to AT&T's competitors and was perceived as predatory by many at the FCC. It finally expired after surviving one of the longest and most complex bouts of litigation in the history of the American jurisprudence system. It put generations of attorneys' children through school.

It was called telpak.

Telpak began its endlessly litigious twenty-year life with the filing of the AT&T tariff, clearly intended to compete with private microwave, in January 1961. It had four service classifications, designated from A to D, equivalent to 12, 24, 60, and 240 voice channels, the space equal to that needed to handle that many simultaneous telephone calls. Also provided would be channelizing equipment to subdivide the "pipes" into channels of lesser width, and channel terminals so that the capacity could be used for telephone calls, much narrower teletypewriter circuits, various kinds of data transmission at differing bandwidths, and telephotograph purposes.

In other words, a customer could add up his communications needs over each leg of a private line network, and then buy whatever size telpak service would meet those needs. If, between two points, the customer could use the equivalent of 150 voice channels, it was still to his economic advantage to buy the 240-voice channel telpak D. If the next leg of the route was lighter, and he was using 45 voice-channel equivalents, he could buy the 60-voice channel telpak C and come out ahead. Besides, if he later needed to increase his use over the route segments, except for channel terminal equipment, the additional channels would be "free," since he was already paying for the larger capacity telpaks.

This came about because telpak pricing was structured to follow the way the telephone system was put together, in channel groups and supergroups. Unit cost of building and operating the telephone system dropped sharply per circuit as the route density increased. Reflecting this trend, telpak A, with 12-voice channel capacity, was priced at $15 a mile under the tariff, but telpak D, with 240 channels, or twenty times as many, would cost the customer $45 a mile, or only three times as much. And even telpak A, since the user was getting a group of 12 voice channels, was competitive in price with the telephone companies' regular private line services.

AT&T's argument in favor of telpak was that it would benefit all telephone customers by retaining and increasing the business of the big users. Heavy private line (telpak) use over a route would mean that more and more AT&T capacity would have to be installed, reducing the cost per circuit mile and the revenue requirements to be met by users of all services.

Wounded competitors, led by the largest manufacturer of private microwave systems, Motorola, and AT&T's only common carrier private line competitor of the time, Western Union, immediately showered the FCC with complaints. After some heated preliminaries, hearings finally began near the end of the year and continued on and off through much of 1962.

The preliminaries included one element which was to be cruicial for the remainder of telpak's life span. As urged by the opponents, the FCC quickly interpreted the tariff to mean that telpak really had to be a pipe, or discrete facility. All channels would have to be derived from a common communication pathway, rather than from different facilities and diverse physical routes.

AT&T and the telpak customers vigorously contested this. They contended everyone was benefiting if telpak were simply considered to be a volume pricing plan, rather than a system of providing service, and it made no difference how the channels were derived. The telephone system and all its customers would gain advantage by the added volume. This would have the effect of increasing route density and reducing per-circuit costs, even though the channels were not all in the same pipe.

When the Commission stuck by its guns, AT&T submitted a tariff change to negate the effect of the FCC's ruling and to spell out its customers' wishes. The revised tariff provided for furnishing of telpak derived channels by diverse facilities. The new provision sharply posed the gut issue of discrimination. Could telpak be distinguished from ordinary private line service, or did it merely constitute a volume offering of like services at discount prices?

The Commission did all it was allowed to do under the law. It suspended the carrier-intitiated tariff modification for the maximum allowable period of ninety days and began a formal investigation of the subject.

At the end of ninety days, the tariff changes to make telpak essentially a pricing plan went into effect. The Commission did not complete the ensuing investigation for years, although the basic nature of telpak as a pricing plan for volume users was now accepted. For the remainder of telpak's long life, mostly spent in court, the central issue was whether its rates could be justified on valid cost or other economic grounds.

The hearings were concluded in mid-September 1962, and for the next year and a half, staff members and commissioners tried to arrive at a consensus when not dealing with other pressing matters. Meanwhile, the big users happily taught themselves how to make the best use of the communications bonanza. Their dreams took on a nightmarish quality, however, in March 1964, when the Commission issued a tentative decision.

Telpak was in fact discriminatory, the FCC said. The services were simply private line at different prices, the agency decided, since it could find no material cost difference in furnishing a given number of channels to one customer as opposed to providing the same number to several users under the private line tariffs. Competitive necessity would justify discriminatory rates if they produced revenues to benefit all customers, it ruled. But it started telpak A and B on the way to oblivion with a conclusion that those classifications were not justified by competitive necessity, since the rates for any given number of channels under the regular private line tariffs were reasonably competitive with private microwave systems of capacity equivalent to telpak A or B.

Nevertheless, the FCC did not shoot Santa Claus for the bigger customers. Since the decision was tentative, they would have another crack at justifying telpak, and the tariff would stay as it was for the time being. In any event, the agency concluded there was "apparent" competitive necessity for the telpak C and D, although it did not have the information to decide whether those classifications were compensatory to AT&T.

The lawyers took over. They prepared voluminous legal documents supporting their clients' positions and jockeyed for position in the oral argument the FCC announced it would hold. Some with common interests reluctantly agreed on one spokeman, but when the schedule was set, nineteen different legal counsel were to appear.

Two days before Christmas 1964, the Commission issued its decision. It left its tentative conclusions much as they were. But recognizing the upheaval that would result, the FCC kept the telpak tariff in place until September 1 of the following year. AT&T was told to "unify" its ordinary private line and telpak A and B rates, in essence washing out the two smaller telpak classifications. It was told to submit cost data so the Commission could decide whether telpak C and D were compensatory and therefore lawful, even if discriminatory.

The following August, the U.S. Court of Appeals for the D.C. Circuit gave telpak the first of the many court reprieves to follow. It said telpak should go on as it was until the court issued a decision on the many appeals which had flooded in. The presiding judge of the panel, later to become more widely known, was Warren E. Burger.

More than a year later, in September 1966, the court upheld most of the FCC's conclusions. But it included strong language citing the desirability of something like telpak C and D for large users, and, most significantly, it said continuance of C and D was not at issue in the case it had decided. The Supreme Court refused to review the decision the next March, and the users found themselves still with telpak C and D, but under a cloud as a result of the nagging unresolved issue before the FCC: Were telpak C and D rates compensatory?

All along, the two sides had disputed how to measure whether telpak was compensatory and thus provided an advantage to all users. The FCC and telpak's opponents supported the fully distributed cost, or FDC, method. Since under FDC all expenses and capital investments are split among the company's services on the basis of how much plant is actually occupied by each one, telpak could have no cost advantage. Because telpak rates for a unit of capacity were lower than ordinary private lines, there was no way under FDC it could earn the authorized reasonable rate of return on investment that all regulated common carriers had the opportunity to achieve. Thus, it would be a burden on the company's other services and, therefore, discriminatory.

Those wishing to retain telpak C and D took an entirely different view. They looked on telpak as an additional service furnished by a company whose basic costs were already covered, provided to customers who otherwise would be "do-it-yourselfers." Without telpak, the business just would not be there. They supported the long-range incremental or marginal cost theory under which telpak would be charged only the additional costs it was directly responsible for. Under their approach, all of the basic costs were being covered, with or without telpak. Thus, if telpak paid the costs it directly caused, and added a profit margin to contribute to the overall enterprise, those users of other services paying the basic costs would receive some help and were that much further ahead.

The sympathies of the FCC and its staff leaned heavily toward the FDC pricing for several reasons. There was the perennnial concern for the financial viability of Western Union, which would have a strengthened position in the private line market if AT&T's competitive service prices were higher. Second, the use of FDC went unchallenged in the earlier AT&T private line investigation, during the 1950s. Finally, incremental or marginal costing was highly controversial in terms of choosing appropriate principles and methods of applying them.

The dispute became even more critical during the ongoing FCC proceedings when, in September 1965, AT&T presented the results of its "seven-way cost study." At the request of the FCC, costs and operating results were developed on the FDC basis for each of AT&T's seven major categories of interstate services. The request came as a result of Western Union's contentions that AT&T had been selectively underpricing services, particularly TWX and private line telegraph, that were competitive with the offerings of Western Union.

On the FDC basis, the study confirmed Western Union's fears. AT&T monopoly services, principally long distance telephone, were earning above 10 percent. TWX and private line telegraph services were producing at a rate of less than 4 percent. Telpak, with its strong economic lure for present and prospective private line customers, showed nearly negative earnings when calculated by the FDC methodology. Private line telephone services provided by AT&T came in well short of AT&T's overall return of about 7.5 percent.

Regulatory implications of the study could not be ignored by the Commission. If the results were taken at face value, public users of monopoly long distance services were overpaying and thereby subsidizing specialized services used predominantly by business. The impact on Western Union's viability was equally obvious.

The long-running battle which was critical to the communications future of large American businesses thus was joined.

In early February 1968, AT&T, looking for the right price to keep both customers and regulators reasonably happy, came in with new telpak C and D rates. They added an estimated $100 million a year to telpak revenues. Telpak C would cost $30 instead of $25 for 60-voice channel capacity, and telpak D would be $85 instead of $45 for 240 channels. The rates were still lower than regular private lines, leaving the discrimination issue. After a ninety-day suspension, the new rates went into effect September 1. Efforts by the big users to get a court stay were unsuccessful.

Unsatisfied, telpak's opponents took off in another direction which was to guide the remainder of its life. If the pricing scheme was equitable and economical, they said, groups of users other than the privileged government and ARINC should be allowed to make shared use of it. Any group of customers should be permitted to buy a telpak service and divide it among themselves.

To AT&T, telpak sharing just would not work. If Tom, Dick, and Harry, all small users, could combine requirements and profit individually from the low rates, they were simply taking advantage of the pricing scheme. There was no chance that they would band together and put up a private microwave system; AT&T considered them just sharpshooters trying to find a price loophole. Later, AT&T was to say that the record of the telpak sharing case was "clear, persuasive, unambiguous, and uncontradicted that a requirement of unlimited sharing would mean the end of the telpak offering." There was not much of an argument about that, and ten years later it turned out to be correct.

More than a year earlier, FCC Common Carrier Bureau Chief Bernard Strassburg had issued a recommended decision concluding that under the law, the telpak sharing regulations were discriminatory. As the FCC looked at its congressional mandate, he concluded, it could not rule that sharing permitted only for the government, ARINC, and a few others was legal. Besides, he ruled, "an offering of private line service which permits broad consumer choice, recognizes cost relationships between the various dimensions of service, and responds to market characteristics should not require the complex tariff structure that has become so difficult to administer and understand."

Strassburg had concluded that the telpak sharing rules were discriminatory, but left it to AT&T to decide how to remove the discrimination. In June 1970, the FCC, upon review of the recommended decision, went much further. It ruled in favor of, and directed AT&T to permit, unlimited telpak sharing. To AT&T's argument that this meant the end of telpak, the FCC—perhaps with a sign of relief—took a "so be it" attitude.

After some further legal manuevering, there began a decade of proceedings in which the federal courts, urged on by the large users, took control. In April 1971, the U.S. Court of Appeals for the Second Circuit (Manhattan) stayed the FCC order requiring unlimited sharing. In late July after argument, the same court ruled with the Commission that the present telpak sharing rules were unlawfully discriminatory, but held that the FCC was in error in concluding on the basis of the record before it that the only right answer was unlimited sharing. It sent the case back to the FCC for another look at the situation.

Not unusual for telpak, court review had been proceeding in parallel in two separate U.S. courts of appeal. The new, higher rates had been taken by the users to the court in Washington. Just before the Manhattan tribunal issued its decision on sharing, the D.C. court upheld the FCC in connection with the rates. It said, in the almost boiler-plate language federal courts often used to affirm regulatory agencies, that the FCC had acted "in the reasonable exercise of its discretion."

September 1971 was another big month for the large users. ARINC started a series of meetings by its board to take a hard, specific look at a private microwave system. Planning some coaxial cable lines as well, it etimated the cost of a nationwide system at $257 million but said that with the new telpak rates its own network would save the airline industry $50 million a year. With the tariff changes, ARINC reported, the airlines' cost of private line service had doubled in seventeen months, from less than $30 million to nearly $60 million, and projections were that the figure might be $140 million by 1980.

Late in the month, the ARINC board authorized the start of planning for the initial two segments of its prospective private system. During the same week, the General Electric Co. said it would launch detailed design of a private system in New England, with planned completion in 1974. As things developed, neither was built.

While all that was going on, the FCC, in response to the court's remand of its sharing decision, told the common carriers offering telpak (basically, the Bell companies, but also Western Union as a reluctant "mirror") to file tariffs in a month, ending telpak sharing discrimination. The Commission concluded that the appeals court gave it the option of leaving to the carriers just how the discrimination would be ended.

To end the telpak sharing discrimination, AT&T essentially had two options. It could open the flood gates to everyone, and that meant discontinuing telpak altogether. Or it could end sharing completely, cutting off the government and regulated entities' access to shared telpak, so that there was no discrimination because no one was allowed to share. Not surprisingly, it chose simply to end telpak sharing.

Telpak sharing, except for the airlines through ARINC, came to an end in December 1971 when the Manhattan appeals court turned down a stay request, pressed mostly by ground transportation entities. The next year, the court issued a decision "on the merits" in the case. It told the litigants, somewhat optimistically, that the solution to the whole program lay at the FCC, rather than in the courts.

The appellants' remedy, the court said, should come at the Commission in challenging at the FCC the tariff provisions which made the airlines a single customer, rather than in fighting to reinstate the sharing provisions which had generally been found to be unlawful. The appellants decided to pass up the challenge.

The die was cast, although it took a number of years before the large users finally lost the telpak battle. In the process, however, they won the volume discount war. Throughout the 1970s, while the big customers struggled to save telpak, two other main streams of events were taking place.

First, the conflict over fully distributed and incremental cost pricing plans, with all of their variations and offshoots, went on. The basic issue arose with all private line services, not just telpak.

Second, AT&T still faced the challenge from private microwave systems, but now it also had to respond to private line service offerings being made by the new specialized common carriers under the competitive policies being fostered by the FCC.

AT&T tried repeatedly with new offerings to meet service demands as user sophistication increased, seeking to find something which would satisfy the regulators while better positioning itself in the competitive environment which was casting an increasingly big shadow. The FCC kept rejecting them as not cost-justified and unreasonably discriminatory. Because customers had to have the services, rejection meant leaving the old tariff—usually already found by the FCC to violate the law after a lengthy investigation—in place while AT&T tried to comply with a new order.

The Commission, mired in a complex proceeding and seeking to find a mutually satisfactory way out of the morass, looked in different directions. A conference of parties' economists, in the hope that there were some economic principles they could agree on, ended after a shower of words. A long-running series of closed-door sessions aimed at establishing some general ratemaking principles did not pan out.

After a further wait of many months, Common Carrier Bureau Chief Walter R. Hinchman, who had succeeded Strassburg, achieved what many had thought was impossible. He was able to produce a recommended decision in the private line/ratemaking principles case, sticking to the FDC method. Four and a half years later, in June 1980, it gained the endorsement of the Court of Appeals in Washington.

Long before that court action, two actions taken by the FCC assured the ultimate demise of telpak. First, as will again be noted in chapter 10, the Commission, in July 1976, adopted FDC and rejected marginal or incremental costs as the basis of pricing AT&T services. In the same decision, it concluded that telpak C and D could not be justified by competitive necessity and that they were therefore unlawful and should be discontinued.

At about the same time, in a separate proceeding, the agency adopted a policy favoring the unlimited sharing and resale of all private line services, including telpak. The Commission explained that it expected this to force carriers toward more cost-based pricing of their services, thus serving the public interest. Under this scenario, all private line services would be available—directly from the carriers, or from resellers or through sharing arrangements—at the lower telpak rates, whether or not individual customers would normally qualify as bulk communications users.

As we observed, those decisions made it certain, once the long court appeals had run their course, that telpak was at an end. But the big users were still the winners. The new world of telecommunications competition and deregulation was moving away from FDC.

The FCC, making decisions with an eye on the courts, Congress, and the industry, was carving out a new era in which AT&T's competitors were more and more being allowed to "let the marketplace decide." The tight regulatory grip on AT&T was being eased somewhat and more sympathy was being directed toward its outcries of "competitive necessity." In that world, fully distributed costs increasingly ran counter to increasing competition and declining regulation.

A few months before telpak came to an end, the government's General Services Administration pointed the way to the big users' answer. GSA administered the Federal Telecommunications System, using 60,000 circuits nationwide provided mostly under telpak by AT&T, over which government employees made official calls (by privately gathered evidence, quite a few were personal in nature). In March 1981, GSA reported plans for competitive procurement. In the first such step, 8,000 heavily used circuits between 200 pairs of cities nationwide were thrown open to the lowest bidder.

The big users' lawyers had done their job. In its final four years, under one of the many court orders, telpak had been "grandfathered," and limited to the existing customers, to spare them harm while the case was being finally decided. Now, the Supreme Court refused to hear the last of the telpak appeals, and the U.S. Court of Appeals for the District of Columbia lifted the grandfathering stay.

At 12:01 a.m., May 7, 1981, telpak, which had made many U.S. industries big users of communications services, breathed its last.

9

"CATS AND DOGS"

The seeds that were to grow until the huge trees they produced choked off a century-old telephone industry structure were planted in the early 1960s. As so often happens, they germinated and poked through the soil without being treated like much more than annoyances. To the telephone industry, they looked very much like weeds, to be treated as such.

Neither planter was a revolutionary. Both had very limited goals. But no one thought Tom Carter or Jack Goeken would leave the nation's telephone establishment (companies, regulators, and customers) changed beyond recognition. Neither of them had the intention of doing so. Their ambitions were more modest; they simply wanted to get a minuscule piece of the equipment and service markets monopolized by the old-fashioned telephone industry. The biggest problem they shared was to get anyone to take them seriously.

In appearance and personality, they could hardly have been more opposite. Carter, well-groomed and soft-spoken, reportedly had some profitable real estate investments in the Dallas area to sustain him as he went about his radio engineering business. His interest lay in mobile communications necessary to support operations in Texas oil fields.

Goeken, rumpled and somewhat overweight, looked and acted the born promoter that he was. He never ceased having brainstorms, whether they were ways to find new forums for his ideas or just to use someone's duplicating machine for necessary documents because he could not afford to have the job done commercially.

They had a few qualities in common—characteristics that were to have important parts in the outcomes of their stories. Both were "nice guys," generating the sympathies of those who felt nonetheless that the two were embarked on useless missions. Neither showed the kind of business acumen needed to profit hugely from their pioneering, although Goeken wound up with a substantial chunk of MCI stock. Most significantly, both had the stubborn determination required to push their projects past large—seemingly overwhelming—obstacles.

The routes they took were as different as their personalities. But again, they had one common thread. Totally convinced that the telephone colossus had nothing to fear from their modest ambitions, they sought, and usually found, every conceivable forum to spread their word. Carter's area of concentration was the private mobile communications field, and no meeting of those interested in what was known as the "safety and special radio" field was too small or remote for him. Goeken knew his answer lay in Washington, and when he was not on the road seeking financing or prospective customers, he was an eager participant in congressional small-business-related hearings.

Neither ever missed a chance for a mention of any kind. Goeken, in particular, was a familiar sight in Washington news offices. He would rush in, sweat pouring off him in the humid Washington summers, with an announcement of some new proposal, usually just in time for a publication's deadline. However erratic and eccentric he appeared, he made it his business to learn things like deadlines.

In the broad sweep of national communications policy, neither seemed to have any unique, let alone earth-shaking, product. Carter worked with oil drillers who had private mobile radio systems connecting the various installations in the oil fields. Often, he observed, there was a need to call someone not within range of the mobile system dispatcher, perhaps to order a part or call for repairs. Or it might be necessary to reach the foreman at his home in some emergency.

To Carter, the solution was simple. Just link the mobile radio system and the telephone network together, by acoustical or inductive connection. Nothing would be hardwired into the telephone system. The Carterfone was, technologically, a simple device—merely an acoustic coupler connecting the telephone in the dispatcher's shack to the mobile radio transmitter.

Carter was surprised when the telephone company told him he just could not do that. It conflicted with the end-to-end service responsibility of the telephone company and could be harmful to telephone service. In the final analysis, it violated the tariffs banning foreign attachments. If oil field personnel wanted to make a telephone call, let them go to a telephone. Efforts to market the Carterfone met vigorous telephone company objections.

Carter's route to the FCC was a little more circuitous than Goeken's. In federal court in Texas, he filed one of the early antitrust suits against the Bell System. The judge found himself on uncharted ground. Where, a decade later, courts readily took on trials of antitrust suits involving telephone issues and found plenty of precedent for them, the Texas jurist had no desire to delve into such unfamiliar areas.

After the preliminary proceedings in the case gave him a good idea of what was involved, the judge invoked the "doctrine of primary jurisdiction." This dispute involving the Communications Act and telephone company tariffs, as well as antitrust law, was a proper subject for the expert agency, he concluded. He referred it to the FCC. After the Commission decided who was right on issues of communications policy, the judge reasoned, he could hand down a ruling under the antitrust laws.

Specifically, the FCC was asked to rule on the justness and reasonableness of the telephone companies' foreign attachment tariff regulations, interpreted as banning connection of the Carterfone to their facilities. The District Court's invocation of the doctrine of primary jurisdiction survived a trip to the U.S. Court of Appeals, and in mid-1966, the FCC's future calendar included the Carterfone case.

Although more direct, Goeken's efforts to obtain an FCC hearing took a little longer. A radio serviceman in Springfield, Ill., he, along with some associates in the radio repair business, looked for a way to generate more volume. Unencumbered by a lot of formal engineering training, and guided by the simple desire to make something work without a great deal of trappings, Goeken slowly formulated what became Microwave Communications, Inc., and later MCI.

MCI, in its beginnings, was a lot like the farming area from which it sprang. If a farmer's tractor broke down when he was hurrying to get a crop in, he hardly had time to wait for a repairman from town. He used the most direct way, and whatever parts were on hand, to get it running again. He was not looking for anything fancy. He just wanted something that worked, and preferably did not cost much.

To a farmer, and to Jack Goeken, things like redundancy and standbys and network control and alternate routes—the heart and soul of the nationwide telephone network, which was always there, performing under any conditions—simply did not exist. A farmer did not buy two tractors so he would have a spare on hand. The Bell System almost always had two ways, and usually more, to reach a given point. To Goeken, that was "goldplating." To the Bell System, it was end-to-end service responsibility.

As Goeken's concept developed, he found prospective customers interested in a direct, simple, "cheap and dirty" private line communications system. Bell System engineers had long since established 4 kilocycles, now kiloHertz, as the optimum bandwidth for a voice telephone circuit. The callers not only heard all the words but the intonations and could recognize the voice at the other end.

It was, they would agree, possible to catch nearly all the words in a conversation over a 2-kilocycle voice circuit. But the sound was distorted and "tinny." A talk over 4 kilocycles was about like a face-to-face conversation. Over 2 kilocycles, one had to strain constantly to make out what was being said.

No matter, some prospective users told Goeken. If he could provide half the bandwidth at half the price, because he could crowd twice as many circuits within a given channel, they were willing to strain a little. Over a private system, they knew they were talking to the St. Louis manager, and what difference did it make if his voice sounded peculiar?

Goeken had some other price-cutting ideas, like part-time and shared private line service at reduced cost. His Chicago-St. Louis system would be based on low-priced billing and built cheaply enough to make a profit on the smaller revenues.

When AT&T built a microwave tower, its lawyers made certain it had a clear title to the property. Goeken would enter into an informal arrangement with a farmer to use a small part of his pasture. AT&T would first build access roads, so repair vehicles could reach the site in any weather. Goeken figured his crew could get their boots muddy in a storm. AT&T engineered its sites so they would work according to established principles and built its structures to withstand the highest velocity winds on record in the area. Goeken thought that if something went wrong, that was farmer's luck, and it could be fixed when conditions improved.

That was the difference between them. That was why AT&T could not understand Goeken, and why Goeken did not comprehend why he was being opposed. There was, in addition, something equally fundamental. To Goeken and his customers, the telephone company would always be there. If something went wrong, his customers had an alternative until MCI was back in service. To the telephone company, later known as the "carrier of the last resort," the very fact that it occupied that role was due to its status as a regulated monopoly. It was there because competitors were not permitted as a matter of public policy and public interest.

The FCC had little experience with "cheap and dirty" telephone service, but it had one precedent which would have given little comfort to Goeken had he known about it. When Western Union acquired Postal Telegraph in 1943, one thing that came along with the deal was a Postal private line telephone system. The voice service was provided via carrier frequencies on the Postal telegraph lines and necessarily was limited to already-connected private line telegraph customers, numbering but a few thousand.

Western Union had no interest in the telephone service: As early as 1879, it perceived no future in voice communication and agreed to sell to the Bell

System its telephone patent rights and operations. It would have taken more than a crystal ball to forecast that, come 1980, Western Union would be back with a minimally successful effort to join other common carriers in their assault on AT&T's long distance market.

In 1950, Western Union and AT&T presented a proposal to the FCC. WU would sell the Postal long distance telephone operations to AT&T, would acquire some remaining message telegraph operations of Pacific Telephone, and would gain a much-needed cash payment of $2.4 million.

In the existing environment, there appeared to be no problem about FCC approval. But the Justice Department had pending its first antitrust suit against the Bell System. It intervened in the FCC case, in favor of the existing if inconsequential competition in long distance service which existed. It never seemed during the hearings that the Justice counsel, Lambert S. O'Malley, had much real support from his front office, but he tried.

At oral argument before the FCC in December 1950, reactions from the commissioners on the bench indicated that the pending application was no pushover for approval. The argument was a throwback to some which had been held on the local level over proposals by the telephone company to replace switchboard operators, who were intimately familiar with their subscribers, with impersonal automatic dial service. Because there were so few customers, opponents of the transaction argued, a Western Union telephone user did not need telephone numbers. He could tell the operator, "Get me Joe in Wichita."

Not liking the way things were going, Western Union's counsel, William Wendt, decided to reverse the situation when his turn came for rebuttal. He suggested that the FCC members visualize what would happen, under the same service conditions, if Western Union had come to the Commission for authority to start a new service. The service was so poor technically, the evidence at the hearing suggested, that calls over 300 miles could hardly be heard. Estimates were it would cost $1.7 million to put the system into decent shape, and meanwhile it was losing $200,000 a year. The number of subscribers was 1,845, down from 2,559 a year before.

"Does anyone want this service?" Wendt asked the Commission, heaping scorn on his adversaries. "Well, there's Joe in Wichita."

The FCC quickly approved the transaction. Wendt was a hero to his colleagues at the always cash-short Western Union, happily picking up the $2.4 million. What it might have been worth to Western Union if it had stayed in the long distance telephone business until the courts and the FCC decided the service was no longer to be a protected monopoly is something that simply cannot be calculated.

For several years prior to early 1967, when both MCI and Carterfone hearings were held by the Commission, the MCI case had been drifting along.

Goeken was not known for being well organized. That probably accounted for some of the delays, while he was modifying his applications or was out rounding up requested information.

His straitened financial circumstances were well known, and his counsel, Michael Bader—who after all had to make a living and could not spend all his time working for a client with no money—sometimes asked postponements because of other commitments. The FCC and the opposing carriers were in no hurry and did nothing to discourage the postponements. At one point, granting a delay, examiner Herbert Sharfman disgustedly voiced the hope that the case would be completed within the lifetimes of the participants.

Finally, everything was about ready. The MCI hearing, starting in February 1967, ran for twenty-four days of sessions spread over nine weeks. The Carterfone hearing, starting and ending in April of the same year, was wrapped up in less than seven days on the record.

The MCI application was vigorously opposed not only by AT&T but by Western Union and the local telephone companies who participated with AT&T in providing private line service. They presented witnesses who emphasized the importance of nationwide rate averaging, under which in essence the high revenues for heavy-traffic routes subsidized operations over light-volume routes. Interstate service between Chicago and St. Louis costs the customer the same as it does between two communities the same distance apart in Montana and North Dakota. To permit a cut-rate competitor to serve a route between two big cities, they argued, was "cream-skimming" which would force established carriers to reduce their rates between the same points and thus endanger nationwide rate averaging.

AT&T could hardly be faulted for not treating either case as the landmark that only later hindsight would justify. Goeken reported that he was looking for a 10 to 12 percent return on an investment of about $500,000. MCI had raised about $151,000 from sales of stock, exclusive of an $85,000 escrow account which would be disbursed only if it won the case.

At one point, the hearings were suspended when Goeken could not promptly produce financial information and leases for his intended tower sites. At another point, MCI's secretary-treasurer, Thomas J. Hermes, a Rand-McNally Co. official, took the Fifth Amendment on his counsel's advice while being questioned about solicitations of prospective investors. A research firm reported a 95 percent probability that the system, if constructed, would have between 58 and 204 customers, and Goeken estimated 35 users would be the break-even point.

The Carterfone case, as befitting one referred to the FCC by a court, centered on the legal status of the tariffs and their interpretation. AT&T stood on the testimony of its vice president of operations, Hubert L. Kertz, that connection of the Carterfone "should not be and is not permitted by

the tariff regulations under which interstate and foreign message toll
telephone services are furnished.'' General Telephone of the Southwest was
also a defendant and took a similar position.

> *Henck's comment:* At *Telecommunications Reports*, we reflected the
> view of our news sources that neither case was very important. Our main
> problem was finding someone on our small staff with enough time to
> cover what we considered rather insignificant hearings. Along with a
> few other minor cases going on at the time, the Carterfone and MCI
> hearings were referred to generically in the office as ''cats and dogs.''

Nonetheless, the potential significance of both cases had not escaped
everyone in the FCC Common Carrier Bureau. Strassburg, as chief of the
organization, was particularly cognizant of the Carterfone controversy. He
well recalled the court decision and the outcome of the Hush-A-Phone case,
as recounted in chapter 4, and had directed the staff working on the upcoming proposed findings and conclusions to take a broader view of the issues
as involving more than merely the Carterfone device itself.

> *Strassburg's comment:* MCI was a little different matter. There was a
> lot else going on, and what I had heard of the hearings was mostly
> negative. Moreover, based on very few personal encounters with Goeken,
> I could not help but question his credentials to contruct, own, and
> operate a common carrier system to compete with the likes of AT&T or
> Western Union. Certainly, he did not fit my stereotype of a successful
> entrepreneur. The case had been dragging on for a long time, and I
> didn't consider it a high priority item. A week before the proposed find-
> ings were due, I initiated getting together with two of the staff econo-
> mists who had worked on the case, Manley Irwin and Bill Melody, and
> discussed at some length what the case was all about with them.
>   Notwithstanding some personal misgivings, it looked like here was a
> legitimate opportunity to factor in a new supplier. The conventional
> thinking was still that the Bell System was a natural monopoly, but some
> changes in the demand side of the environment were beginning to show
> themselves. We were starting to think about how to accommodate data
> transmission on the network, and we had started the computer inquiry
> which exposed some new dimensions of the communications infrastruc-
> ture. Other options were being examined, with the emphasis on simply
> trying to fill in the voids, rather than confronting the Bell System eyeball
> to eyeball.
>   Jack Goeken's Chicago-St. Louis proposal looked like a safe and pru-
> dent guinea pig. I couldn't see any harm to be done by giving it a try to
> see what happened. We would find out whether there was market for
> the specialized services he was proposing. So I issued instructions to
> rewrite the bureau's position, which had been drafted as a denial, to
> recommend a grant.

Still, wondering how the commissioners would react, I told the staff to make it clear we were not proposing a commitment to a new policy, but rather to put the applications in a "test tube" motif. Earlier in the same year, consistent with past policy, the Commission had turned down a GTE application to build its own microwave system in the Pacific Northwest to handle toll traffic. The reasoning was that AT&T had adequate facilities to provide the service.

The bureau's proposed findings, recommending a grant to MCI, came as something of a bombshell. Reflecting Strassburg's thinking, the document said the MCI proposal "should be given a reasonable opportunity to become established and thereby afford the Commission an opportunity to observe the results and consequences of its operations in terms of the demand generated for its services, its ability to operate efficiently and profitably, and its effects upon [competing carriers'] operations."

The case arose, the bureau found, "because of the impact of advancing technology upon consumer demands for communications services and the structure of the communications market. Advancing technology permits the establishment of new services that will meet consumer needs and demands that previously had not been satisfied. . . . Commission policy should endeavor to bring about market conditions that will facilitate the application of new technology and the full development of untapped markets."

Rejecting the established carriers' cream-skimming arguments, the bureau concluded, "Virtually all of the service lost by the existing carriers to MCI is likely to take place because the characteristics of the MCI service offering are more attuned to the specific needs of a submarket of communications users. . . . If MCI is successful in its endeavors, the size of the total communications market would be expanded. In successfully developing this potential submarket, MCI's service offering would tend to stimulate use of complementary communications services provided by the existing common carriers." Furthermore, it was argued, the submarket would be so small that it would not cause any significant disadvantage to the larger carriers.

As time is measured in government circles, it did not take long for Examiner Sharfman to issue his initial decision. In October 1967, less than four months after the proposed findings and conclusions were filed, he held that grant of the MCI applications "may not be an unalloyed blessing," and in fact "they may be invitations to disaster," but "MCI should be given an opportunity to show that it can compete productively."

The projected system "is not gold-plated," he conceded, but there "is no reason to believe [it] will not work." The nationwide rate averaging scheme, he said, is "embodied neither in the Decalogue nor in the Constitution. Without danger to the republic, there may be a weighing of the possible public benefits or disadvantages resulting from authorizing competition in selected areas.''

Sharfman's decision reached the 1980s a decade and a half early. He commented further:

Clearly, if the averaging doctrine is sacrosanct, and "cream-skimming" is an attendant horrific to the extent the carriers claim, they have insulated themselves against private line competition except from carriers with unlikely initial operations as widespread as theirs. They are like courtiers who deign to accept challenges for duels only from those of equal rank. But the efforts of a relatively impoverished newcomer, proposing a novel if by no means faultless service, to give battle on his chosen ground should not be impaled by this agency upon a principle devised by his opponent.

Finally, he reflected the view of MCI which just about everyone held at the time. He commented that the "MCI sites are small; the architecture of the huts is late Sears Roebuck toolshed, and they are without the amenities to which Bell employees, for instance, are accustomed, and servicing and maintenance are almost improvisational; but there is no reason to believe that the system will not work, unless one is bemused by the unlikely occurence of the catastrophes to which the carrier witnesses gloomily testified.''

> *Henck's comment:* Back in the days when FCC lawyers were given special temporary assignments as hearing officers to conduct specific cases, a practice widespread in the government, a wise man once said to me: "You watch those guys. They will keep edging forward one step at a time. Eventually, they will be wearing robes, and they will have all the trappings of office, including bailiffs."
>
> He was not far from being completely right. Now, they are known as "administrative law judges," they are called "Your Honor," and everyone rises when they enter the room. When the FCC built the present hearing room arrangement, they no longer had to mingle with the common people during recesses. They slip in through a private entrance, just like a real judge in a courtroom.
>
> Herb Sharfman accepted the future, in many ways, in the most calm and unassuming manner imaginable. I once told him, during a hallway conversation during a recess, about the prediction that had been made to me. He replied, "Oh, that wouldn't be so bad, Fred. Then I could wear my old gardening clothes to work, and cover them up with a robe."

It took nearly two years for the FCC to take final action in the MCI case. Things moved more quickly where the parallel Carterfone proceeding was concerned.

Less surprisingly than in the MCI docket, the Common Carrier Bureau recommended that the tariff provisions generally limiting use of customer-provided equipment be cancelled. The bureau, going back to the language of the Hush-A-Phone court decision, urged a tariff provision that "clearly and affirmatively states . . . that customer-provided equipment, apparatus, circuits, or devices may be attached or connected to the telephones furnished by the telephone company in the message toll telephone service for any purpose that is privately beneficial to the customer and not publicly detrimental."

The bureau took the position that the telephone companies should be required to set forth in their tariffs "reasonable standards and requirements that any such customer-provided equipment . . . must meet in order to protect telephone company employees, facilities, the telephone system, and the public from adverse effects."

There were, in the bureau's findings, two main points which became lost in later developments. First, the bureau was against any arrangement under which customer-owned equipment would be substituted for what the telephone companies provided when they furnished end-to-end telephone service. What was proposed as permissible was the connection or attachment of devices to "telephones furnished by the telephone company."

Second, the use of protective devices to safeguard the telephone system was accepted. That had been the consistent policy line over the years. At this stage, neither the bureau nor any other party was challenging this policy of protection.

A short time later, Examiner Chester F. Naumowicz, Jr., issued his initial decision. Instead of the broad policy change proposed by the bureau, he stuck strictly to the Carterfone itself. Finding that there was no showing the device would harm the telephone system, he commented, "We here consider only a specific device and the evidence as to what, if any, effect it will have on the system."

Naumowicz made one other finding which, if it had been upheld by the FCC, would have spared AT&T some later grief in the antitrust courts and probably would have saved its shareowners some money. He ruled that the tariffs had been properly construed in the past to bar the use of the Carterfone and, therefore, were not unlawful until a specific finding was made that an attachment would cause no harm and should be authorized for future use.

Under FCC rules, exceptions to Naumowicz's initial decision were due in forty days. As usual, they were filed by all interested parties who did not agree with any part of his decision. The Justice Department was heard from for the first time, saying it did not intervene before because it would have been duplicating the bureau's efforts. Now, it supported the "broad relief" urged by the FCC staff, describing it as not only responsive to the Communications Act "but to the broad policy of antitrust laws."

Opening up the network to new types of equipment obviously was much more universally popular at the FCC than new competition in intercity transmission. Taking only about one-third of the time it did to decide the MCI case, the commissioners accepted generally the bureau's Carterfone position by a unanimous vote in late June 1968.

The Commission concluded in effect that the interstate telephone system was wide open for interconnection to any private system through any device which "does not adversely affect the telephone company's operations or the telephone system's utility for others." It ordered that two sections of AT&T's message toll service tariff, containing the general prohibitions against interconnection and foreign attachments, be stricken. The order gave the carriers permission to propose new tariffs "which will protect the telephone system against harmful devices," and said they "may specify technical standards if they wish."

At AT&T, the decision did not lead to the cries of alarm which often accompany the loss of what was now—if it had not been considered so earlier—a major case. Company officials appeared concerned only about one aspect of the FCC decision. The Commission had ruled that the tariff section governing connection of prohibited devices in effect for more than ten years had been unjust and unreasonable for the entire period. The tariff provision had been unlawful since the Hush-A-Phone decision, the FCC stated, even though it had accepted the tariff language following the Hush-A-Phone case.

AT&T correctly assumed that the FCC ruling on past unlawfulness would lead to numerous damage claims and asked the Commission to reconsider that point. Otherwise, company statements reflected the tone of a speech made by Chairman H. I. Romnes earlier in the year. AT&T's top official reported that efforts were under way to open the network and accommodate new types of transmission, particularly data.

The initial AT&T statement, after announcements of the FCC's Carterfone decision, was this: "We have had the whole subject of connection of customer-owned devices in our network under intensive review for some time. Our intent is to be as responsive as possible to evolving communication needs through making our network available to expanding uses. At the same time, safeguards are essential to assure that other users of the network are not adversely affected."

A short time later, the company advised the FCC that it was "well along" in formulating new tariff regulations which "will provide for the connection of customer-provided terminal devices through service-protective arrangements." They would be broader than that, AT&T observed, in requesting a little more time. "The matter of general interconnection of customer-provided communications systems involves serious and complex problems which go beyond those involved with the connection of technical devices."

The new tariffs were filed some weeks later, after extensive discussions with increasingly delighted FCC officials. As will be recounted more extensively in chapter 11, they went well beyond the strict or literal requirements of the Commission's Carterfone decision.

For all practical purposes, the Carterfone case itself was over. There was only a brief skirmish at the U.S. Court of Appeals in New York. AT&T and the General Telephone System filed brief protective notices of appeal, to safeguard their legal positions, but withdrew them when the FCC accepted the new AT&T tariffs.

What the Carterfone decision would lead to had scarcely begun. It remains as one of the relatively few seminal rulings in a half century of policy change. As things stood near the end of 1968, noted in *Telecommunications Reports*: "After more than 40 years, the secluded couple in Vincent Youmans' 'Tea for Two,' who refused to have it known that they owned a telephone, can come out of hiding, provided the telephone is an extension off a private switchboard."

Back at the federal court in Texas, soon after the new tariffs took effect, the eight-year-old triple damage civil antitrust suit of the Carter Communications Corp. against AT&T, Southwestern Bell, and the General Telephone Co. of the Southwest was quickly settled for a reported $375,000. Initially, the plaintiffs had asked $1,350,000.

A little more than a year later, without giving reasons, Carter announced his "complete resignation" from the company. He would remain the company's largest individual stockholder, it was said, and would enter the consulting field.

Meanwhile, near the end of 1968, well over a year after Sharfman's initial decision recommending the grant of MCI, there were well-authenticated reports that the Commission was split 3 to 3 in the MCI case. A new appointee, Democrat H. Rex Lee, reportedly was still studying the matter, and it was indicated he could be the swing vote in the decision. As it turned out, he was.

After the fall election of 1968, speculation was rife in the telecommunications industry as to how appointments to the FCC by newly elected President Richard Nixon would affect the MCI case. As events developed, the Commission lineup was unchanged when the decision was rendered.

Finally, in mid-August 1969, the FCC issued the decision. By a 4 to 3 vote, it granted MCI's application to provide specialized private line services between Chicago and St. Louis. The vote was an unusual one at the Commission, along strict party lines. Although Nixon had been in office for seven months, none of the Democrats' terms had expired, and he was to be unable to establish the four-member Republican majority permitted under the law for another year.

The majority termed the case a "very close" one, but it bought Sharfman's reasoning, in somewhat less colorful language. "If we were to follow the [opposing] carriers' reasoning and specify as a prerequisite to the establishment of a new common carrier service that it be so widespread to permit cost averaging, we would in effect restrict the entry of new licensees into the common carrier field to a few large companies which are capable of serving the entire nation. Such an approach is both unrealistic and inconsistent with the public interest."

A sharply divergent view of the FCC's public interest standard came from the dissenters. They described the decision as "diametrically opposed to sound economics and regulatory principles," and one that was "designed to cost the average American ratepayer money to the immediate benefit of a few with special interests."

It was noted at the time that local and short-haul interconnection would lead to further controversy, as it did. The Commission said it would retain jurisdiction over the question of interconnection, since the established carriers "have indicated that they will not voluntarily provide loop service" needed by MCI to connect its customers to the terminals of its microwave system.

The following January, the same four commissioners joined again as a majority as the FCC turned down petitions that it reconsider the ruling. This time the margin was 4 to 1, as two of the Republicans had left the agency, and the two new members did not vote. The concept of what MCI was proposing remained unchanged.

Once again, the decision was that "the availability of a two kilocycle voice channel for those not having a need for a nominal 4 kc voice channel and the part-time sharing and flexibility provisions of the MCI proposal will make the cost of the communications service lower for the subscriber who desires to take advantage of this service." There was support in the record, it was stated, for the conclusion that "a public need exists for the part-time and channel sharing provisions of MCI's tariffs."

Once the MCI decision became public, something akin to the Oklahoma land rush began. Within several months, there was a flood of MCI-type applications for other high density parts of the country. In one of their opposition statements, the opposing carriers pointed out that "no longer can MCI-type applications be regarded as an isolated experiment."

A large number of the applications came from the MCI organization. Goeken had no interest for long in being a business adminstrator, and William G. McGowan, a successful business consultant and entrepreneur, moved in to take over the management. A new Microwave Communications of America was set up and began to disgorge applications for hundreds of microwave stations crossing high-volume regions, such as New York-Chicago, the West Coast, and Twin Cities-Chicago.

History now records that on the day Bill McGowan arrived on the job, the limited vision of MCI's future held by its founding father, Goeken, and by the FCC majority began to be replaced by a much broader one. He was joined within the next year by Kenneth A. Cox, who became MCI senior vice president and of counsel to Mike Bader's law firm. Cox's term had expired at the FCC, where he was one of the four-member majority in the MCI decisions.

What happened next is a story for future chapters. But one conclusion was obvious at the moment the Commission released its MCI decision. As noted in *Telecommunications Reports*, "The domestic telecommunications business will never again be the same."

10

THE FCC GOES PUBLIC

During its first thirty years, the FCC never seriously attempted to conduct, let alone complete, a formal on-the-record investigation of the Bell System's rates for interstate message toll telephone service. On occasion, it would hold a public hearing on the rates for some of the ancillary services, such as private line. But the big ticket, bread-and-butter service—message toll telephone—escaped public scrutiny by the FCC, which preferred the informal nonpublic approach of "continuing surveillance."

In October 1965, however, the FCC, to the delight of its staff and the dismay of AT&T, did an about-face by plunging the rates for all of AT&T's interstate services into the public arena of a comprehensive, formal investigation and hearing.

The commissioners were told by their staff that the "fundamental and far-reaching" issues before it, directly involving Bell System rates, related equally to the entire domestic common carrier industry. The Common Carrier Bureau advised the FCC members:

> The nature of the issues is such that it is not possible to resolve them through either informal discussion or application of the process of "continuing surveillance."
>
> They involve complex issues of fact and policy regarding which interested parties—carriers, users, and state regulatory bodies—have diverse if not diametrically opposed views. Accordingly, these regulatory, policy, and factual issues urgently require that appropriate formal proceedings be instituted without delay in order to make possible a meaningful resolution of the issues on the basis of a public record.

The people walking the FCC corridors expected that customers would get more via the formal hearing route. But the choice was not so much one of more or less. The Commission was faced with some old, nagging questions now grown much sharper with new information, and the day had arrived to try to answer them. Other long-pressing issues had developed more of a cutting edge simply because they had been around so long.

The bottom line issues had to do, of course, with the quantity of AT&T's interstate earnings. There was, first, the ever-present dilemma of what the authorized interstate rate of return should be. And then came: applied to what? Basically, that issue revolved around three contested items totaling about $700 million in the Bell System's interstate rate base: telephone plant under construction but not yet in revenue-producing service; cash working capital (the question here was not whether to include any, but how much was the minimum needed to operate the business and who supplied it, the investor or the ratepayer); and materials and supplies on the shelf to keep the wheels turning, but not necessarily part of the capital investment which comprised the rate base.

Also permeating those bedrock issues were the concerns as to whether all of the multibillions of dollars entered into plant and operating expense accounts represented prudent and necessary outlays properly recoverable from the customers of the Bell System. How much, if any, should be disallowed as the product of waste, inefficiency, overstaffing or overmanagement? After all, as a monopoly, Bell had few of the conventional incentives at least theoretically present in competitive industry to maximize efficiency.

And as previously noted, there was the ever-present but elusive issue that gnawed at the regulatory conscience. Most of the telephone plant and equipment that made up the Bell System network was manufactured by Western Electric Company, the then almost 100 percent-owned (and later totally owned) AT&T subsidiary. What Western did not manufacture, it purchased from the general trade, which was excluded from having any direct dealings with the Bell operating companies.

Western's prices were generally lower than those prices paid for comparable items of equipment manufactured outside of the Bell System for the independent telephone industry. But Western's volumes were vastly greater. If its prices were not as low as they could be (obviously, an extremely difficult question to try to answer), then Western was a source of unjustified and hidden extra profit to its parent, AT&T, and this extra profit found its way into the rate base of the operating telephone companies. Ratepayers, in turn, would be paying too much for service because AT&T's authorized rate of return would be applied to an artificially inflated rate base.

As the FCC staff made clear in its landmark October 1965 memorandum to the Commission, determination of a proper level for the Bell Sytem total interstate earnings would not resolve all pressing telecommunications regulatory and economic problems by any stretch of the imagination.

In a day when the crisis-prone Western Union was virtually the only competition to the Bell System as well as the only provider of message telegrams, its continued existence and viability were matters of considerable regulatory concern. (It might be observed that the desirability of having a modest competitor in existence was not totally lost on AT&T. AT&T sold

WU its competing teletypewriter exchange (TWX) service on relatively favorable terms under only mild pressure. More important, it kept WU afloat for years—intentionally or not, that was the result—by leasing WU its backbone facilities under contracts which have been later demonstrated to provide minimal rates, in relation to their costs. The exchange of facilities contracts was not terminated—they ran indefinitely, subject to termination notice by either party—until MCI and other competitors were pressing AT&T hard for equal treatment with Western Union, and it could not justify giving WU lower rates than the other companies.)

The FCC a few years earlier, in 1963, had started a domestic telegraph investigation to make policy judgements as to how to deal with the problems and prospects of that industry. As was noted previously in connection with the private line rate case and the telpak proceeding, WU's diagnosis of its ills and its prescription for a cure were fairly simple. It felt that its troubles stemmed from AT&T's use of huge revenues from its profitable monopoly message toll business to subsidize too-low rates for AT&T's services which competed directly with WU. The cure was for the FCC to stop it.

FCC staff members contended that a crucial part of the problem was that earnings of the AT&T services directly competing with WU's basic services constituted too much of a moving target. The Commission would order an investigation of the rates of one of those services—telpak, TWX, private line—and AT&T would carve out figures covering only the service under investigation. Why not, they asked, insure across-the-board accounting for *all* of AT&T's total costs and revenues by dividing them up in pie-chart fashion so that when figures for all the services were compiled, they added up to 100 percent?

Developing ground rules for the "seven-way cost study" took a while. The project's title came from the seven major AT&T service groupings: message toll, wide area telephone service (WATS), TWX, private line telephone grade, private line telegraph grade, telpak, and all other (mostly program service to radio and TV stations). The way the pie was to be cut among them amounted to the big issue.

We have recounted in chapter 8 the basis for the fully distributed cost (FDC) versus marginal and incremental cost approaches to this kind of study. As pointed out, FDC is a relatively simple approach compared with the others when it comes to determining whether rates for particular services are reasonable and nondiscriminatory. It relies on average historical costs as recorded on the books and a distribution of those costs based on the extent to which the plant has been actually occupied by each of the several services during a past period. It is no accident that the method used in the first seven-way cost study is now referred to as "pure" FDC.

Pricing a service on the basis of FDC necessarily results in higher rates than would result from the use of marginal or incremental costs, for a mix of reasons. FDC pricing therefore makes it more difficult to retain or attract the business of large users who find competitive alternatives—such as contructing their own systems or taking service from other suppliers—more economically attractive. But marginal or incremental costs are by far more difficult to calculate, involving controversial economic theorems and a large measure of subjectiveness.

The results of the first seven-way cost study were so sharply skewed that, even using the rate base most favorable to AT&T, rates of return for the twelve months ended August 31, 1964, ranged from 10.2 percent for WATS to 0.3 percent for telpak. In between, on the company's basis, message toll was 10 percent, while TWX was 2.9 percent and telegraph grade private line was 1.4 percent. All of this occurred within an overall AT&T interstate rate return of 7.5 percent. On the FCC staff's basis, of course, the results were even further over all of the landscape.

It would have been fairly hard to make sufficient rate adjustments to allay any concern about cross-subsidy. It was difficult to avoid at least some small suspicion that the effect of AT&T pricing practices—planned or not—was to use profits from the then-monopoly message toll and WATS services to pay the freight for services competitive with Westen Union. WU then had to match those low rates. Cost-study methods were sufficiently in dispute that they did not provide a detailed road map for rate adjustments. But they underscored the need for regulatory concern.

For the FCC to engage in a full-fledged on-the-record examination of Bell System interstate rates, costs, and earnings, it could not avoid the one subject that was the point of departure for all rate regulation, whether by the FCC or the states. That subject was jurisdictional separations. The states had just recently given their approval to the interim use of revised separations procedures—the Denver plan—and were awaiting FCC concurrence, since no firm rate decisions could be made without a system of separations mutually acceptable to the federal and state jurisdictions.

The FCC had agreed to the use of the plan, but not without misgivings as to the rationale offered in its support—a rationale fashioned by AT&T, which had devised the plan at the behest of the states. More important, the FCC was concerned with the kind of precedent the new plan would set in this complex area of separations.

The Denver plan represented the first decisive break with the principle of relative or actual use as the sole measure for allocating the subscriber line portion of exchange plant costs to the interstate jurisdiction. The FCC felt some sympathy for the argument that had been advanced for some years that relative or actual use (some 4 percent to 5 percent of total use) made by interstate calling of subscriber line plant was an unrealistic measure of the value of that plant to the interstate long distance network. After all, subscriber line plant was not traffic sensitive. It was equally available for local and long distance calling, and its costs did not vary with the volume or nature of the calls transmitted over that plant.

Nothwithstanding the FCC's empathy with the states, its major over-riding concern was that departure from the jurisdictional allocator of relative or actual use would open up separations to anyone's subjective judgment as to what allocation produced fair and equitable results. Relative use as the sole allocator may have been unfair, but its virtue lay in its manageability and relative certainty.

Put all the considerations together, including the criticism of FCC's regulatory performance, which had been coming sporadically but disturbingly from Capitol Hill and other sources, and it was clear that to the Commission they spelled "formal rate case." The agency put behind it thirty-one years of negotiated AT&T rates. It came close to putting behind it forever the continuing surveillance process, although the closed-door talks were repeated once, four years later, under special circumstances.

To no one's surprise, AT&T was not terribly happy about the unanimous FCC decision. In fact, Chairman Fred Kappel sought to avert the action. He paid agitated visits to each of the commissioners, admonishing them against what he considered a rash and unproductive move. The visits backfired. Commissioners who may have had some reservations about the advisability of a general investigation now were strengthened in their resolve.

AT&T also filed a petition for reconsideration, and its top officials made speeches declaring that the demands on managment of an all-out rate case would impair the Bell System's ability to serve the public. It looked to the investment community as though the FCC was after AT&T's scalp, and the market price of the company's stock declined sharply.

Behind AT&T's reaction was its long-held and probably correct belief that the FCC staff was more liberal (translated, in this instance, as more inclined to lower rates) than the commissioners. Over the months in which procedures for the case were being established, AT&T sought to have the FCC staff separated from decisional considerations at the agency. Its contention was that the agency's prosecutorial functions should be divorced from its decision-making, or judicial, activities. The argument on the other side was that the commissioners should not be deprived of the advice of the FCC's best-informed staff members.

Eventually, a compromise was reached, which has continued today, in the rare instances when the FCC resorts to formal public hearings instead of receiving written comments and evidence. A separate trial staff was set up, and FCC employees assigned to it were directed not to discuss the case with decision-making personnel. Instead, the trial staff filed recommendations like any other party.

By no means, however, was the Common Carrier Bureau's front office separated from the decision-making function. In fact, as the procedural framework has developed in the years since then, the bureau has been delegated increasing authority to take actions in its own name, sometimes in areas of major significance. The chief of the Common Carrier Bureau remains, on a day-to-day basis, the government's most important official dealing with telecommunications. It is just that no one fights it any more.

At the close of 1965, the FCC reaffirmed its commitment to the formal hearing, turning down the petitions that it reconsider its basic order. Responding to the procedural objections raised by AT&T, FCC Chairman E. William Henry, in a separate statement, summarized the Commission's position.

The order, Bill Henry said, "simply reflects a concern that certain matters must be considered, at this time, on a public record. . . . Seldom if ever before in the Commission's history has there occurred such an extraordinary confluence of regulatory problems in the field of common carrier communications. . . . Another problem now at high tide concerns the validity of the so-called separations formula. . . . These questions require something more than the usual roundtable discussion or informal hearing."

The Commission was off on what turned out to be an eighteen-month ground-breaking exercise in the "gut" part of the proceeding, which became known as phase 1-a. The first session, a prehearing conference, held January 31, 1966, turned out to be a memorable one. The nation's capital was literally paralyzed after a two-foot snowstorm, and the government was officially closed, but presiding Commissioner Rosel H. Hyde (noting that in his native Idaho the storm would have been considered routine) went ahead with the conference. With some having displayed considerable ingenuity in travel, twenty-seven people were in the room, for an event which otherwise would have drawn at least a hundred. Ironically, when phase 1-a closed more than a year later, on an ordinary winter day, sixteen people were on hand.

Transcript of phase 1-a covered more than 10,000 pages, not including a two-foot shelf of "canned" direct testimony not copied into the hearing record. The FCC commissioners, a few months later, were given more of an education than some may have wished. They heard a two-day oral argument in which the first long day was devoted entirely to separations, and the second to not-much-easier-to-understand considerations of rate of return, capital structure, and controverted rate base items.

By invitation of the FCC, three "cooperating" state commissioners had sat in on the case. They had no vote, but were free to voice opinions. Ostensibly, this gave the FCC the state-side perspective. Two went along with the AT&T request for earnings in the 7.5 to 8.5 percent range, while the third opted for the Commission's eventual conclusion that 7 to 7.5 percent would be good enough. They were in agreement that the contested rate base items should be retained. The first two supported the Bell System's capital structure, while the third was in favor of a suggestion by one of the witnesses put on by the FCC staff that AT&T should be considered to have 45 percent debt capital for rate-of-return purposes whether it did so or not.

During most of the years that the consolidated Bell System existed, AT&T's capital structure was a prime point of regulatory dispute. Companies could obtain debt capital, like bonds and debentures, at rates appreciably lower than they could sell new equity (stock). Thus, the more debt in the company's capital structure, the lower the total cost of money.

Conservative-minded AT&T urged that anything over 40 percent of its total capital would increase its degree of risk in the eyes of the investment community. This in turn would trigger a warning signal among bond-rating companies, whose ratings are a critical factor in how much bond money costs a borrower, and cause an increase in its total cost of capital. Regulators viewed their rate-paying constituency as ahead of the game at a 50 percent ratio and cited as examples the electric utilities, which generally maintained even higher debt ratios which they handled without evidence of significant risk to their bond ratings or overall cost of capital.

When the hearings were held, AT&T's top financial officer, Jack Scanlon, argued strongly that AT&T would not have been able to raise capital on suitable terms during inflationary times if it had not had the borrowing capacity of a conservative debt ratio.

Later events showed in effect that both sides were right. During the inflationary periods of much of the 1970s, AT&T could not sell new stock at the current dividend and growth rates without offering so much of a premium that existing shareowners would dump their holdings. Thus, it resorted to somewhat lower cost debt financing, and the debt ratio climbed very close to 50 percent. So yes, AT&T proved it could handle a 50 percent debt ratio. And yes, the road would have been much rougher without that borrowing margin made possible by a lower debt ratio.

The FCC's phase 1-a decision in early July 1967 limited itself to an expression of support for a debt ratio above 40 percent. The Commission thought this would be in everyone's best interest, but did not go as far as the harder-nosed regulators urged and impute a higher figure in computing the allowable rate of return. In any event, AT&T's Scanlon announced from the witness stand his company's plan to increase its debt ratio above 40 percent.

In areas of much more immediate concern to the contending parties, the FCC directed an interstate rate reduction of $120 million. It based its bottom-line figures on the conclusion that a rate of return in the range of 7 to 7.5 percent was reasonable, and that on the basis of its rate base findings AT&T had earned 8.56 percent in the preceding year, which the FCC used as its test period. All of the contested rate base items were disallowed.

The rate reduction would have meant earnings of about 7.75 percent on the new rate base, but another half percentage point would be cut as the effect of the FCC's prescription of its own separations formula in place of the Denver plan, which the agency rejected.

Like all major government decisions based on judgmental exercise, and involving rapidly changing circumstances and the interplay of economic forces, the FCC's rulings did not stand unchanged for long. AT&T started the ball rolling with a petition for reconsideration, asking that the rate cutback be reduced to $75 million because the FCC's rate base rulings made its allowance of a 9 percent return on AT&T equity "mathematically impossible."

Examining the current business scene with sharpened pencils, the Commission showed some willingness to yield a bit in a decision issued very quickly thereafter. Because of the rate base changes and the uncertainties stemming from effects of the new separations plan and its concomitant impact on settlements with the independent telephone companies, it deferred $20 million of the $120 million rate reduction to the following April 1. It reversed itself on plant under construction, $544 million of the $700 million total which had been knocked out of the rate base.

During the entire period of phase 1-a preliminaries and hearings, the separations pot had been boiling. The states acted at times as though they were about to be thrown into the pot to provide the main course for the FCC's dinner, while the latter viewed its mission merely as one of trying to regain a measure of control over what it regarded as the illogical results of the Denver plan.

The several years of unusually heated controversy and trauma which followed add up to one of the clearest examples of difficulties of divided regulation of the combined Bell System, with its control of the principal local companies in all of the forty-eight contiguous states along with the nationwide interstate telephone network.

Naturally, the FCC had to be concerned primarily with interstate rates, while the state regulators were basically responsible to the makers of local and intrastate calls. Both were dealing with a commonly owned national telephone system whose costs and revenues had to be divided among jurisdictions on the basis of the sometimes-arbitrary separations procedures, and with single local telephone systems employed by telephone users for both intrastate and interstate calls.

To recount in detail the rate/separations activities of the next several years would tax readers' patience and their ability to maintain a focus on constantly changing sets of numbers, as the closely interested parties argued, negotiated, and proposed revised separations plans. The membership of one informal committee of experts trying to reach some compromise agreement showed that the closely interested participants included not only the FCC, AT&T, and the state commissions' association, the NARUC, but also Western Union, then AT&T's major competitor, and the federal government's General Services Administration, representing the largest user of telephone service.

All efforts to come to an accord were in vain by the time the Commission reached its phase 1-a decision in July 1967. To cut the knot, the Commission exercised its "best judgment" and formally prescribed a separations plan for its own jurisdiction. It was completely different from any of those proposed by the parties, and it wiped out the Denver plan.

Most significantly, the FCC for the first time departed from its past rigid adherence to relative use in allocating the joint costs of subscriber line plant between exchange and interstate services. The plan, in simple terms, called for the restructuring of the interstate allocator by adding a weighting factor to the actual use measurement. It was intended to reflect the fact that toll rate schedules are based on time and distance and therefore act as a deterrent for long distance calling compared to most flat rate exchange service rate schedules, which present no such deterrent for local calling. By weighting actual interstate use, the FCC was seeking to make allowance for those deterrents.

Finally, the FCC's action was the first time that it "prescribed" the separations procedures. This made them subject to change only by formal public rulemaking, rather than by federal/state/industry off-the-record horsetrading which had determined past revisions.

In its decision the FCC sought to bury, once and for all time, the states' exploitation of the state/interstate toll rate disparity problem. The FCC commented, "We do not regard the mitigation or elimination of disparity between state and interstate toll rates as a valid consideration in determining the propriety of separations procedures. We stress that the purpose of such procedures is to determine the amount of investment, expenses, taxes, and reserves subject to federal and state jurisdiction. Their purpose is not to effect an artificial or contrived equalization of the cost per unit of service among all jurisdictions."

Disparities, the FCC went on, are "a consequence of the multiplicity of regulatory jurisdictions over message toll services. Not only are different ratemaking philosophies and practices applied in each of those jurisdictions, but the structure of costs and patterns of usage differ from state to state and between the federal and each state jurisdiction."

The separations situation stayed fluid, until it overflowed in November 1969. In spite of the rate reductions the FCC had ordered less than two years earlier in phase 1-a, interstate earnings continued to press upwards. So once again, enticed by the prospect of a relatively quick and easy interstate rate reduction, the FCC returned for the last time to the continuing surveillance approach when dealing with the consolidated Bell System. After a series of closed meetings, the smiling FCC reported—to a public including an angry group of state commissioners—what was then the largest interstate rate reduction in history.

An interstate message toll service (MTS) rate reduction of $237 million, the Commission said, should not result in earnings going below the 8 to 8.5 percent range. First, it stated, MTS rates would go down $150 million on January 1, 1970. The next month, charges would drop another $87 million, but to AT&T these would be offset by recently filed rate increases for program transmission, telpak, and TWX services, as a result of the ongoing "phase 1-b" proceedings in the interstate rate case. To produce the $150 million, AT&T proposed at the beginning of December substantial reductions in customer-dialed station calls at any time, plus a new one-minute base rate for station-to-station calls after midnight. The rate cutbacks submitted to be effective February 1 heavily emphasized the wide area telephone service.

Thus, once more the FCC, in order to trim interstate earnings, resorted to rate reductions rather than a change in jurisdictional separations to benefit the states. NARUC had been trying for some time to have the Communications Act amended to require the FCC to convene federal/state joint boards to act on separations and other matters which impacted the regulatory authority of the states. The FCC interstate rate reduction announcement, with no accompanying changes in jurisdictional separations, triggered an immediate demand for full-scale congressional hearings on a joint board amendment.

FCC Chairman Dean Burch saw no reason in the record for the Commission to reach a purely political compromise on what he viewed as the "minimum contraints of law and reason which must govern in any formulation of jurisdictional separations if it is to pass judicial review." Burch, an Arizonan who was a principal manager of Barry Goldwater's 1964 Presidential campaign—in the days when he was a heavy smoker, he was a prototype "Marlboro man"—had a rocky and often difficult relationship with the state regulators.

As grass-roots office holders, the state commissioners, acting in concert, had considerable cumulative political weight on those rare occasions when they exercised it in unison and in a strong states' rights emotional upsurge. This time, they did. When the Senate Commerce Committee opened hearings in mid-December, ostensibly on a joint board amendment to the Communications Act but more broadly on federal/state relationships in general, representatives of thirty-two states packed the hearing room. At the same time, NARUC formally presented to the FCC a petition for rulemaking to approve a transfer to interstate of $250 million in revenue requirements through a change in the allocation of local dial switching equipment.

(The mills often grind slowly where separations is concerned. Local dial switching equipment has remained a subject of controversy, as the next step in the telephone system's service hierarchy beyond the constantly chewed-over non-traffic sensitive or subscriber line plant. Seventeen years later, a joint board was still considering the possibility of separations changes affecting local dial switching equipment.)

Members of the Senate and House greeted the states' complaints with sympathy and encouraging comments, but the complexities of common carrier communications remained a near impossible subject for comprehensive legislation then, as now. The proposed joint board legislation did not escape the legislative mill until the FCC's Strassburg and NARUC's general counsel, Paul Rodgers, hammered out a compromise on procedures. The agreed-on rules for a separations joint board were very similar to those later codified into law by Congress as a new section 410(c) of the Communications Act.

Under the then-negotiated plan, and now the law, the states have a 4 to 3 edge in joint board membership. The board, however, issues only a preliminary proposed decision. The final verdict comes from the FCC, during deliberations in which the state commissioners can join, and speak their piece, but cannot vote. (When procedures were set up, the FCC had seven members, and it was theoretically possible for the federal joint board members to be ultimately outvoted. Now, with five FCC commissioners, the three federal joint board members automatically constitute a majority of the final decision-making body. The state commissioners may have a majority on the joint board, but they are politicians. And like all politicians, they can count. The result is a remarkable similarity in decisions of the joint boards and, later, the FCC.)

The outcome of some five years of separations travail and trauma, going back to the short-lived Denver plan, would stay in place just about as long as the consolidated Bell System. In August 1970, after NARUC meetings at Lake of the Ozarks, Mo., the joint board approved what was memorialized as the Ozark plan. The FCC gave its approval two months later. It was initially calculated to transfer $120 million in annual revenue requirements to interstate, but caused increasing problems over the years due to its tendency to escalate beyond expectations in the percentage of total non-traffice sensitive costs being transferred to the interstate side.

For the next generation or two, the Ozark plan of separations may be best known for establishment of the subscriber plant factor, known in print as SPF and in conversation as "spif." The first, relatively simple part of the two-part formula was said to develop "the basic subscriber plant cost of an exchange call and assigns to each interstate holding time minute of use the same cost as is assigned by the formula to an exchange minute of use." The second part is an additive "to recognize the deterrent effect of the interstate toll rate schedule specifically related to interstate calls originating in each study area."

Only an advanced mathematician could explain the algebraic formula which produced the SPF factor. It did bring about the desired result, although subsequent growth of interstate toll volumes caused a shift in benefits to the states which exceeded the expectations of the plan's architects. To correct the situation, SPF was subsequently frozen by application of a transitional phased-in formula at a 25 percent assignment of non-traffic sensitive costs to the interstate jurisdiction.

As far as regular interstate message toll service was concerned, the first formal rate case was practically over. Attention now turned to phase 1-b in which the Commission and affected parties struggled mightily with the ratemaking principles to be employed in deciding on the proper rate and earnings relationships that should exist between message toll, by far the biggest part of the total interstate business, and such specialized services as private line/telpak, TWX, and program transmission. As ever, the controversy centered on cost allocation principles and methods or, more especially, fully distributed versus incremental costs.

In the process, two valiant but fruitless efforts were made to achieve agreement, after it became obvious that the on-the-record testimony was turning into recitations of the special concerns of each interested group. First, the presiding hearing examiner, Arthur Gladstone, convened as an informal part of the hearing a conference of the parties to determine if a stipulation of relevant factors would help produce a statement of ratemaking principles that everyone could live with.

The effort did not succeed, and the next move was to get the parties—carriers, competitors, users, and regulators—to try to work out their differences by bargaining in closed, off-the-record sessions.

At the end, after a couple of months, the principal outcome was a set of very general principles. Beyond paying lip service to the relevance of fully distributed and incremental costs under appropriate, but undefined, circumstances, the parties agreed on little else except that the determination of proper costing and pricing would have to be made on a market-by-market basis. It was thus apparent that having put all the ratemaking issues together in one big case, the FCC was now going to have to pull them apart again. The conferees recommended, and the Commission agreed, that issues be added to each of several pending cases to consider the justness and reasonableness of the Bell rates for individual classes of service. The cases concerned telpak and private line, broadcast program transmission services, and TWX, soon to be sold by AT&T to Western Union by agreement of the two companies.

> *Strassburg's comment:* If there was ever a policy issue which gave me an acute sense of ambivalence and indecision, the FDC versus incremental cost allocation question was it. Even though I felt that incremental pricing was defensible in a competitive environment, I feared for the future of that environment, which was still somewhat embryonic, unless AT&T was held to FDC pricing, for two reasons.
>
> First, its application by AT&T could be more readily policed than incremental costing. Second, it would force AT&T to maintain higher prices for competitive services, at least until its competitors had a reasonable opportunity to penetrate the markets.

Splitting up the remaining issues on justness and reasonableness of rates for non-message toll services, the FCC followed the plainly evident course and terminated phase 1-b.

It was not until much later that the vexing costing/pricing controversy, which permeated all of AT&T ratemaking, was resolved with finality. In January 1976, Common Carrier Bureau Chief Walter R. Hinchman, who had succeeded Strassburg two years earlier, achieved what many had thought was impossible. As already covered in chapter 8, he produced a recommended decision in the private line/telepak case prescribing the FDC method as the primary standard to be used in pricing AT&T services. Hinchman's recommendation was adopted by the Commission's decision of the following July. Four and a half years later, in June 1980, it gained the endorsement of the Court of Appeals in Washington.

There now remained phase 2 of the general rate case, whose principal focus was the reasonableness of Western Electric prices and profits on its sales to the Bell operating companies.

However, there was little, if any, enthusiasm in the FCC to engage this question. The staff, in particular, recalling the impasse reached in the 1956 private line proceeding in making any firm judgments as to the reasonableness of Western's prices and profits, anticipated a repeat performance. In the absence of a substantial boost of manpower with which to conduct an independent staff investigation of Western's operations, the staff would again be a prisoner of the evidence offered solely by AT&T on the virtues of Western.

AT&T, on the other hand, was eager to get back on the record with its evidence on Western Electric. On several occasions during 1969, AT&T officials urged Strassburg to activate phase 2. The company, they said, had devoted substantial money and effort to preparing its case and felt convinced that it had a credible, persuasive story to relate concerning Western Electric prices and the contributions that vertical integration made to more efficient and economical telephone service in the United States. Strassburg readily conceded that, in the absence of countervailing evidence, AT&T would inevitably prevail. It was preferable in his judgment for the FCC to do nothing.

Strassburg also told the AT&T officialdom that the Commission was seriously considering an alternative approach to the traditional examination of Western's prices and profits. In fact, the FCC had tentatively agreed, subject to obtaining the minimal funding from the Congress, to embark upon a comprehensive fact-finding inquiry into the entire equipment market structure in the nation.

The inquiry would not have as its prime focus the question of whether Western in the past had overcharged the operating telephone companies of the Bell System. Rather, the inquiry would take a future look by addressing the effects of vertical integration on the costs and efficiency of telephone service, its impact on equipment and service innovation and improvement, whether to impose competitive procurement upon all operating telephone companies, and whether to recommend new legislation to implement the Commission's findings. The FCC never got to its market structure inquiry, but AT&T finally got the opportunity to present its case on the public interest virtues of vertical integration.

As noted earlier in this chapter, the FCC, in November 1969, after resorting for the last time to the continuing surveillance approach, announced that it negotiated with AT&T further reductions in interstate long distance rates of some $237 million annually—an action which aroused the ire of the states and led to another jurisdictional separations change to be effective January 1, 1971.

But within less than a year after the negotiated rate reduction took effect, AT&T concluded that its action was a mistake. It advised the FCC that the rate cut was premised on estimates of future economic conditions that had not materialized and that it had to recoup some $150 million. In addition, it also was required to offset the effects of the separations change. In short, it filed for an increase of $385 million, which it claimed would yield a rate of return of 9.5 percent—two percentage points above the upper part of the 7 to 7.5 percent range which the FCC prescribed in phase 1-a.

After some quick dickering between the FCC and AT&T, the agency committed itself to give AT&T an expedited hearing in which the rate-of-return question would be resolved in advance of all other issues. In return, AT&T reduced the amount of its proposed increase to $250 million.

Included in the material filed by AT&T in support of its proposed rate changes was extensive testimony and data it had carefully crafted but had been prevented by Strassburg from coming forward with in phase 2 of the old rate case. Thus, the FCC would listen to AT&T's recounting of the virtues of vertical integration in the Bell System whether it wanted to or not.

In setting AT&T's new rates for investigation and hearing, the FCC again phased the proceedings into two parts: phase 1 to deal principally with rate of return and rate base, and phase 2 for the old, unresolved issues presented by Western's prices and profits.

But this procedural gesture still did not mean that AT&T was finally to have its day in court to vindicate Western and vertical integration. The proceeding in phase 1 dealing with interstate rate base and the allowable rate of return was concluded in record time. While the proposed decision of the hearing examiner was being evaluated by the parties and the Commission, Strassburg had growing concerns about the imminence and logistics of phase 2, dealing with Western. Staffwise, his bureau was no better equipped to take on AT&T's well-honed case on Western than it was in prior years.

He discussed his problem with Chairman Dean Burch, who was quick to agree that to proceed on the Western phase without the proper expertise would be a sham and disservice to the public in a matter of such fundamental importance. What were the alternatives available under these circumstances? The FCC could indefinitely postpone going forward with phase 2, or just dismiss the proceedings until another day when the FCC was more generously endowed with requisite resources. The situation was discussed with the full Commission, and, with a split vote and a holding of breath, phase 2 was ordered dismissed.

An FCC order taking the action pointed to the priority matters now pending before the agency which would intensely affect the using public. They included specialized common carriers, domestic satellite policy, further carrying out Carterfone terminal equipment policies, and ratemaking policies. Strassburg told the press that there was no assurance that the results of phase 2 would lead to any rate changes, while the matters being concentrated on involved "substantial public benefits." In the best tradition of all organization heads, he cited staff and budget shortages and the need to make the best use of what he had.

The reaction from the public, politicians, and consumer groups was immediate and shrill. This was, they felt, an admission by the regulators that they would not or could not regulate. Discussions ensued with those officials of the Executive Branch who held the FCC's purse strings. The promise that funds would be forthcoming enabled the FCC to corral a special staff and to contract for the services of outside specialists to provide the wherewithal for a meaningful examination of the Western Electric issue. In less than a month, the FCC ordered a reinstatement of the proceedings.

The proceedings which followed consumed some 103 days of hearings. Eventually, four and a half years after phase 2 was reestablished, a 535-page initial decision was issued by the presiding officer, David I. Kraushaar. He concluded that the structure of the Bell System, and for that matter the structure of its rates, should remain unchanged.

In February 1977, at the time when all concerned would have been startled to learn that in less than five years an agreement to dismember the Bell System would be announced, the FCC finished off phase 2 and the overall rate investigation docket. It unanimously ruled that the overall service performance of the Bell System was "excellent" and that the record had not produced anything to call for a fundamental change in its structure.

The rate proceedings had served their purpose. Throughout the 1970s, and into the start of the next decade, inflation laid a heavy hand on all kinds of costs, in particular for regulators the cost of money. The rules of the road laid down by the big rate case enabled the FCC to deal with constantly pressing telephone rate issues. As investors made it evident they were insisting on much higher returns, the authorized interstate rate of return went up several times, finally to the 12.75 percent level where it stayed until three years after divestiture. Telephone rates increased, inevitably. But, though the huge volume of business meant that even nominal increases produced frighteningly large totals, the percentage contribution of telephone rates to the skyrocketing cost-of-living index was small in comparison with other expenses.

When the FCC reinstituted phase 2, after an increase was provided for the purpose in President Nixon's budget proposal, comments on the scene included those of Commissioner Nicholas Johnson. Nick Johnson was usually a minority of one on the FCC, a constant dissenter and critic of large business enterprise. He took the occasion to write a separate concurring opinion, voicing grave doubts that the action would do anything for the causes he espoused.

Johnson spoke warmly of the rates and service flexibility of goverment-operated European telephone systems, and said Europeans would "consider it preposterous greed for the rich to profit while poor consumers pay for the much higher capital costs involved in raising billions of dollars by the sale of private stock, rather than public or private bonds or other funds." But he saw little hope in the FCC's track record for the kind of public service structure he advocated.

His final words in the January 1972 statement, just about ten years before the antitrust consent decree was announced, were: "If we are not to regulate with appropriate resources to the job, the alternatives of nationalization or breaking up AT&T into separate corporations that are more manageable [and competitive] simply must be considered by Congress and the American people."

11

THE END OF END-TO-END

Even though the FCC's landmark Carterfone decision, which we have discussed in chapter 9, went far beyond questions raised solely by Tom Carter's device and cancelled the foreign attachment strictures of the AT&T tariffs, no one at the Commission expected that it would lead to substitution of telephone-company-provided equipment by terminals provided and owned by the customers.

Describing the Commission staff concept, FCC staff spokesman Kelley E. Griffith told the FCC members in oral argument in the Carterfone case, "We are not here concerned with any idea or notion that customers may be able to substitute their own telephone handset for the telephone company handset, or that they put in their own loops, or that they may do anything other than make attachments or connections to the message toll telephone service, as defined and manifested in the ordinary telephone handset."

In the Carterfone decision which followed, the FCC stated that "the facts of this case did not involve the furnishing of purely telephone system equipment telephone-to-telephone on the message toll telephone system."

As Strassburg has written in a review of post-Carterfone and terminal interconnection history, "By this qualification, the FCC drew a boundary line between the customer and the telephone network which barred the customer from substituting any of his own equipment for that furnished by the telephone company in providing its telephone service. This included not only the telephone instrument, but any other customer-provided equipment having a network signaling capability, such as private branch exchanges or key systems which would likewise be beyond the bounds defined by the Carterfone ruling."

The no-substitution point may be the single most misunderstood fact about the Carterfone decision—especially by plaintiffs and sometimes judges and juries in antitrust proceedings—in the ensuing decade or more. To readers now becoming accustomed after five years or so of terminal deregulation to owning or leasing their own telephone sets, and being responsible for the inside wiring beyond the telephone company's connecting block, the logical question then becomes: If the FCC did not initiate the policy of customer-owned terminal equipment in the Carterfone decision, how did it get started?

The answer lies in our comments in the foreword to this book that while most of the crucial rulings affecting telephone service in the past half-century have come from government sources, the responses of industry itself have been an important contributing factor. It lies also in what may have been a long-term if unspoken AT&T policy, or perhaps only an accident of history, that a "hard-line" AT&T chief executive officer was usually succeeded by a "softer line" successor and vice versa.

As the term is used, the "hard-line" AT&T chiefs—in the modern era, Leroy A. Wilson, Fredrick R. Kappel, and John D. deButts—were motivated first and always by the deeply held conviction that it was the Bell System's mission to provide universal end-to-end telephone service as a natural monopoly regulated in the public interest. They fervently believed that this was best for everyone, especially the subscriber, and constituted the Bell System's principal reason for being.

The "softer line" AT&T heads—Cleo F. Craig, H. I. Romnes, and, as it developed, Charles L. Brown—were equally convinced of that tradition and mission. But they were more likely to be pragmatic advocates of compromise when it became apparent that the changing times simply dictated some change in course to carry out these goals rather than undertaking a determined fight beyond the last ditch. (Of those mentioned, Charlie Brown was the only one who found himself being pushed into the last ditch. Those listed as hard liners might well have made the same decision if they had faced the same circumstances.)

"Hi" Romnes's time in AT&T history was not as dramatic or traumatic as Charlie Brown's. But, coming as it did in the wake of the Carterfone controversy, it was a very significant period, to say the least. Questions immediately arose regarding the company's reaction to the FCC Carterfone decision in two areas—whether it would appeal the ruling to court, and what kind of tariffs it would file to replace the cancelled provisions relating to customer-owned equipment.

The Carterfone case was held, and decided, in the midst of a confluence of questions about maintaining end-to-end telephone service under all circumstances. The FCC, noting the blurring of the former clear-cut distinction between voice telephone and data communications service, had started its first computer inquiry.

The purpose of this inquiry was to ferret out and address the regulatory and policy issues that were likely to emerge from the growing convergence and interdependence of the two technologies.

Of no less importance, the inquiry compelled AT&T, all other segments of the telephone industry, and the entire computer establishment to focus on how this interaction would affect their economic and marketing futures. In particular, the centerpiece of concern was the capacity of the voice-oriented telephone to respond effectively to the data communications requirements of the new Information Age. The inquiry elicited a vast amount of information and opinion that could not be ignored by policymakers and planners of both government and industry. Much of the input questioned AT&T's capabilities to continue to be all things to all people in communications.

Some time before the FCC issued the Carterfone decision, Romnes pointed out in a speech that, with the onset of substantial interest in data communications and the possibility of attaching customer-owned data terminals to the telephone network, the Bell System had been looking at ways to "open up" the network.

It was soon obvious that Romnes was trying, with the caution warranted by a drastic change in the environment, to move one careful step at a time into the new world. AT&T, under his leadership, was ready to surrender its complete monopoly of terminal equipment and to make room for a new industry standing poised to rush to the marketplace as soon as the FCC gave it the cue.

Romnes and Strassburg conferred about the Carterfone impact on several occasions, and the FCC official became more enthusiastic at every turn as it developed what AT&T had in mind.

Both company and regulatory commission were convinced of one thing: Every adequate safeguard had to be taken to insure against harm to the quality of telephone service that could result from indiscriminate, uncontrolled hook-up of foreign attachments. The concept of protective connecting arrangements had long been established, originating in 1947 with the FCC's authorization of telephone recorders discussed in chapter 3. So at first no one seriously considered any alternative. Proposals for equipment registration, in which manufacturers would certify to the FCC that their products met minimum safety standards, came along a little later.

With the benefit of 20/20 hindsight, it now can be seen that the Romnes/Strassburg concerns of possible harm to telephone service, and for that matter the people providing it and using it, were overstated or are yet to be realized. This was understandable. A system whose quality and reliability were universal was moving into uncharted waters. People whose qualifications and character were unknown would be vigorously marketing all sorts of gadgets for attachment to telephone lines. Who knew what might happen?

The fears of telephone company management are best exemplified in the later government antitrust trial testimony of Henry M. Boettinger. Hank Boettinger, a brilliant and versatile scholar, was assigned to a small headquarters group as AT&T's "thinker in residence" or, perhaps more accurately, radical in the front office. His job was to develop "far out" alternatives, options, and reactions to events, forcing the top management to consider all the possibilities.

Observing that AT&T's fiercely held conviction about potential harm to the network was not being generally accepted during the years of controversy which followed Carterfone, Boettinger, tongue only half in cheek, came up with a suggestion to prove the point. The proposal, since he picked Indiana as a typical U.S. area, became known as "blowing up Indiana." Boettinger recommended, half seriously, that a trial be conducted in one geographic area of unlimited interconnection of every device customers wished to install, without protective arrangements. At some point, it was expected, the network would become overloaded with trouble and would collapse, thus demonstrating graphically by destructive testing the danger that AT&T was talking about.

Boettinger recalled that AT&T officials, weaned on the sanctity of telephone reliability, were aghast at his proposal. They immediately pointed out that the result would be interruption of service to police and fire departments, doctors' offices, hospitals, and similar locations. There was no way telephone veterans would even admit the possibility of such a cynical step.

The point of Boettinger's story was that there was no doubt in the minds of AT&T management (and this was generally true at the FCC) that a telephone network burdened with all sorts of customer-provided equipment, without protection, would in fact "blow up."

In one aspect, AT&T's reaction to the Carterfone decision was emphatic. The Commission, overturning one of its staff's recommendations, had ruled that the foreign attachment tariffs of the Bell companies had been unlawful since their inception. AT&T asked for Commission reconsideration of this part of the decision, and Romnes made it clear that this element would be appealed if the FCC did not reverse itself. AT&T had no small concern that the FCC finding of retroactive unlawfulness exposed it to a proliferation of private antitrust suits.

Strassburg commented in testimony in a private antitrust case, "In view of the Commission's long history of treating exceptions to the general prohibitions of the foreign attachment tariffs on a case-by-case basis and its acceptance of the tariff revision in 1957 as complying with the Hush-A-Phone decision, I did not regard this retroactive condemnation of the foreign attachment tariffs as warranted or fair."

As noted in chapter 9, the Carterfone case had come to the FCC on remand from a court considering Tom Carter's antitrust complaint against the

Bell companies. The Commission may well have been motivated by the fact that its conclusion on retroactive illegality had the effect of protecting the Carterfone claim for past damages.

AT&T filed a one-sentence notice of appeal with the U.S. Court of Appeals in New York in November 1968, after submitting its wide-ranging post-Carterfone tariffs. After the tariffs became effective, AT&T withdrew the appeal, stating that it had been filed to protect the company's legal position and now was unnecessary because the tariffs had been allowed to become effective. After a brief oral argument, the court granted the motion to dismiss. (This sustained the FCC finding of retroactive unlawfulness of the foreign attachment tariff provisions, a development not especially helpful to AT&T in later defending dozens of antitrust lawsuits by purveyors of various kinds of telephone equipment.)

The company's tariff response to the Carterfone decision was, except to insiders, surprising. AT&T proposed sweeping changes in the tariff rules relating to interconnection of customer-owned equipment. The new tariffs would permit acoustic or inductive connection of customer-owned equipment, including private mobile systems (the function of the Carterfone).

Of greatest significance was the right given to the customer to directly connect devices such as PBXs, key sets, and data modems which were traditionally regarded as an integral part of end-to-end telephone service. However, the revised tariffs specified that the connection of customer equipment and systems was to be made to the telephone network through a "network control signaling unit" and a protective connecting arrangement (PCA). Both were to be furnished, installed, and maintained by the telephone company as part of the company's facilities. The connecting arrangements would be provided at a monthly tariff fee running about $2.

After further discussions with the FCC and other interested organizations, AT&T submitted further "significant and far-reaching changes" in its tariffs, this time relating to customer-provided communications systems. Wide-open interconnection was permitted, except that AT&T insisted on control of network switching and signaling functions through protective devices. Interconnection of private systems would have to be at the customer's premises. This was intended by AT&T to prevent "piece-out" of private networks, by their partial patchwork use of AT&T's facilities.

The AT&T tariffs, as Strassburg had testified,

went well beyond the requirements of Carterfone. They permitted customers not only to connect their own attachments, but also to substitute their own PBXs, key telephone systems, and even telephone sets for those provided by the telephone company. A principal condition specified, however, that the telephone company would retain control of the network control signaling function and provide a connecting arrangement to protect the network from harm.

The revised tariffs were a source of considerable gratification to the Commission and its staff. Those tariffs represented a quantum leap forward in opening up the telephone network to the wider use of customer-provided terminal equipment. At the same time, they could be expected to stimulate the development of a whole new industry selling terminal equipment directly to telephone subscribers.

Immediately before the end-of-the-year holiday season in 1968, the Commission ruled that the entire package of tariff changes should go into effect January 1. In doing so, the FCC rejected a number of protests arguing that the new tariffs failed to comply with the Carterfone ruling in that they did not permit customers to attach equipment of their own which performed network signaling functions. The agency also provided for a series of informal engineering and technical conferences, "broad in scope," to discuss "what further changes are necessary, desirable, and technically feasible."

In another step of considerable importance in later litigation, the FCC—however overjoyed many of its people were with the turn of events—did not prescribe or formally approve the tariffs. Instead, it simply let them go into effect as filed on the scheduled effective date provided by AT&T.

The Common Carrier Bureau, Strassburg later testified, did not recommend formal FCC approval for several reasons:

First, although the tariffs were in compliance with the Carterfone ruling, they would now permit customers to substitute their own terminals for those theretofore provided by the telephone companies as part of telephone service.

In this respect, the new tariffs represented a fundamental change in the parameters of telephone service for which telephone companies were traditionally responsible. We were not certain as to all of the public interest implications of these fundamental changes in the public undertaking of the telephone companies. It was our feeling that this aspect of the new tariffs warranted further consideration.

Second, there were the unresolved questions raised by comments of interested parties with respect to various specific features of the protective connecting arrangements of the tariffs.

And third, there were obvious concerns arising from the dependence of new independent suppliers of terminal equipment upon their competitors—the telephone companies—for connecting arrangements and network signaling units. Consequently, although the new tariffs complied with the Carterfone ruling and, in addition, liberalized customer interconnection beyond the purview or expectations of that ruling, we did not feel that we had a basis for formally approving the tariffs.

Moving into 1969, the first full year after the Carterfone decision and after AT&T's responses changed the face of U.S. telephone service, the FCC went down several parallel avenues. It asked for written comments on the way customer-furnished devices would be integrated into the system, and later it set up advisory groups to offer recommendations on nontechnical aspects of the rules to be followed. At the same time, it arranged with the

National Academy of Sciences (NAS), the most impartial technically oriented organization available for the job, to study technical and scientific aspects of the material submitted. This was the first time that the academy contracted to assist a regulatory agency in this fashion.

NAS had its Computer Sciences and Engineering Board set up a fourteen-member panel to analyze the considerable amount of written material submitted to the FCC. The fact that panel members were not "pure" scientists in the sense that they drew paychecks immediately caused criticism. It was a symbol of changing attitudes that most objections were raised because two of the fourteen panelists were officials of the Bell System. The others were employed by nonprofit and/or government organizations, non-Bell manufacturers, independent telephone companies, or large users of communications services.

Harvard Professor Anthony C. Oettinger, who headed the NAS Computer Sciences and Engineering Board, and the panel chairman, Lewis Billig, Technical Director of the Communications Division of the Massachusetts Institute of Technology-affiliated Mitre Corp., rejected references to the panelists as "representing" anyone. They said panelists were chosen to get a proper "mix" of scientific expertise, and that the only way to avoid a panel with the other "interests"—government or academic as well as business—would be to select a group on the basis of its "innocuousness." The panel had the same membership as originally appointed when it began its deliberations.

The advisory committee convened by the FCC divided into several subcommittees. They began what turned out to be an arduous, lengthy series of meetings in September 1969. That they never produced dramatic, detailed agreements was not the fault of the members, in view of the diverse interests they represented. (Unlike the NAS panel, there never was any doubt in the advisory committee meetings that the members were speaking for their employers, and no one ever suggested otherwise.)

What the committee did produce, in sessions to which the members and their employers contributed very substantial amounts of time, travel, effort, and expense, was consensus—not unanimous but that of a large majority—which became more and more apparent as the meetings went on. Eventually accepted by the FCC in decisions which were ratified by the federal courts, the majority opinion was in favor of equipment registration or certification as the most generally accepted way to protect the telephone network.

Meanwhile, the NAS panel showed a good sense of realism and understanding of the situation as it was taking shape. The group issued its report in about nine months, in June 1970. It concluded that totally uncontrolled interconnection of customer-furnished equipment would be harmful to the network, as it existed at the time, and recommended two possible approaches.

Either or both could be employed, the committee said, and if the two were used in parallel, they could be utilized "in such proportions as non-technical factors" (outside the panel's study area) might determine.

The optional recommendations to supply necessary protection for the network, including the vital element of network control signaling, were: "(1) protective arrangements as required by the tariffs; and (2) a properly authorized program of standardization and properly enforced certification of equipment, installation, and maintenance."

Not surprisingly, the critical nontechnical factors were economic in nature. Making customers hook up devices not provided by the telephone companies by using tariffed protective connecting arrangements (PCAs), the outside suppliers argued, gave the phone firms a distinct competitive advantage. They contended that the PCAs were unnecessary, and that customers looking at an extra $2 monthly cost for each piece of terminal gear they owned and attached would decide in favor of simply leasing the units from the telephone companies. In charging monopolistic intent, they pointed out that the telephone companies provided some outside-manufactured gear to their customers at monthly tariff rates without requiring that a PCA be used in the connection.

There were, in fact, instances when the identical product offered by the telephone companies under tariffs could also be purchased directly by the customer on the open market. Strassburg wrote AT&T that this situation seemed unnecessary and unjust, in that a PCA requirement should not apply to equipment that was available to the general public from the general trade, when the same product was manufactured in accordance with AT&T specifications, purchased by the Bell System, and offered to the subscribers under tariffs with no requirement for a PCA.

AT&T and some state commissions condemned this step by the Common Carrier Bureau as jeopardizing the orderly and uniform implementation of the already liberalized interconnection tariffs and intruding on the work of the advisory committees in the effort to devise alternative approaches to the PCA. Rather than press the issue and rock the already unstable boat of federal-state relationships, the bureau quietly dropped the matter.

As the controversy continued, it became apparent that it could be settled only by formal regulatory intervention and not through any sort of industrywide agreement. In June 1972, the FCC established a federal-state joint board to consider whether, and under what conditions, customers would be permitted to provide their own network control signaling units and connecting arrangements. Essentially, the board was to consider those issues that went beyond the Carterfone decision, providing the forum for formal consideration of the numerous recommendations which were being and had been received from the National Academy, the FCC's advisory committees, and its own chief engineer, among others.

By its order creating the joint board and specifying the issues it was to address, the FCC gave its first public recognition that customers were to have the option to furnish their own network control signaling units, which supervised the setting up and termination of a telephone call, including its timing and billing.

This meant that telephone service as an end-to-end joint venture of the telephone companies was soon to become a thing of the past. It represented the beginning of the regulatory road which the nation has traveled to greater and greater fragmentation of the telephone network and the responsibility for its planning, operation, and maintenance.

AT&T had been asked by the bureau on several occasions to provide statistical data on the incidence of harm to the telephone network. In April 1973, it reported findings that customer-owned equipment attached to the interexchange private line services it was providing "is causing a much higher trouble rate than is experienced when all telephone company provided equipment is utilized." A study of "maintenance of service" reports—charges made when customer-reported troubles were found to be the fault of attached outside equipment—indicated that "nearly 30 % of the troubles caused by customer-provided equipment resulted in harm to the network," AT&T stated.

Five weeks later, the FCC Common Carrier Bureau, after an analysis of the materials it had been furnished, replied that "statistically meaningful differences" of network troubles between customer-provided and telephone-company-furnished equipment had still not been demonstrated.

By March 1974, the outlook had crystallized to the point where Strassburg told a gathering of customer-owned equipment purveyors, the North American Telephone Association founded by Tom Carter, that they "must look elsewhere than to regulation for a solution" of what he now termed a tariff competitive advantage of the telephone companies.

In what may have been the first public hint of what the FCC was to decide six years later in "computer inquiry II," he proposed a restructuring of the Bell System under which "all suppliers of customer equipment and systems would be subject to the same rules and market forces of supply and demand. Thus, the Bell System, if it is to participate fairly for interconnect business, should do so through a separate subsidiary or subsidiaries."

A short time previously, one of the seminal cases in the entire process had its beginnings. The North Carolina Utilities Commission decided to issue a proposed rule designed to nullify the Carterfone decision and its implementation. The rule would have flatly prohibited customer-furnished equipment in the state and called on telephone companies there to own, maintain, and be responsible for all equipment used in intrastate telephone service. It asked for written comments and scheduled a hearing.

As North Carolina moved toward making the proposal final, and as a few other state commissions took actions of similar thrust (such as one in Nebraska which would have required hotels and motels owning their own telephone systems to be regulated as common carriers), the FCC moved in early 1974 to protect the federal jurisdiction. It granted a request for declaratory ruling filed by the North American Telephone Association on behalf of its members, most prominently a North Carolina firm, Telerent Leasing.

Thrust of the FCC ruling was that state commissions could take no action which would reverse or be inconsistent with prior federal rulings relating to interconnection of customer-owned equipment and systems to the "indivisible" telephone network. Otherwise, the Commission held fast to its prior decisions in the area. It also announced that it would issue supplemental instructions to the joint board to look into the economic impact of interconnection and determine if the effects were sufficient to cause reversal of the earlier rulings on public policy grounds.

But the Telerent ruling was of seminal importance in one crucial respect. It was the first of a series of rulings by which the FCC preempted the exercise of state authority in areas of communication that had heretofore been subject to dual federal and state regulation.

The North Carolina commission promptly took the matter to the U.S. Court of Appeals in Richmond, Va. The National Association of Regulatory Utility Commissioners submitted its own appeal, and other states joined on an individual basis.

The federal tribunal took a long time in dealing with the case, for reasons which later became apparent, but no delay was evident in elements of the debate going on in the industry and in other forums. Included was probably the most notable formal address ever made by an AT&T chairman.

AT&T Chairman John D. deButts, never one to keep his strongly held views about AT&T's end-to-end service responsibility a secret, became in 1973 the first AT&T head to address an annual convention of the NARUC in nearly a half-century. When he finished, a confrontation which was to come to a climax in Washington offices and courtrooms nearly a decade later was well on its way.

Throwing down the gauntlet, deButts told the big crowd which came to hear him, "The time has come for a moratorium on further experiments in economics—a moratorium sufficient to permit a systematic evaluation, not merely of whether competition might be feasible in this or that segment of telecommunications, but of the more basic question of the long-term impact of its futher extension on the public at large, the adequacy, dependability, and availability of its service, and the price it will have to pay for it."

Fundamental to the Bell System's approach at the time was the AT&T chairman's keynote that "we cannot live with the deterioration of network performance that would be the inevitable consequence of 'certification' and the proliferation of customer-provided terminals that would ensue from it. No system of certification we can envision—and no interface requirement— can provide a fully adequate alternative to the unequivocal and undivided responsibility for service that the common carrier principle imposes."

Addressing the burgeoning competition both in terminal equipment and intercity service (at the time, limited to private line), he raised this question:

Can there be competition—real competition—when not all the parties to it enjoy the same freedoms or bear the same responsibilities, endure the same contraints? A free market, if it is truly free, affords not only free entry but free exit as well. And competition, if it means anything, means that all the parties to it are equally at liberty to choose which markets they will serve and which they won't. . . .

I doubt that there is anyone in this room who thinks that's going to happen or that it will be permitted to happen.

DeButts's speech is well remembered, but nearly forgotten is the extent to which the bulk of state regulators agreed with him. The principal action by the convention that year—by voice vote, with no audible dissents—was adoption of a resolution calling for efforts to obtain legislation providing a federal surcharge on all interstate communications, both common carrier and private systems, with the proceeds to be distributed to the states to reduce intrastate telephone revenue requirements. All interstate communications, regardless of how sent, would have in essence been taxed, with the fund then disbursed under a formula devised by federal-state joint board to bring down local and intrastate telephone rates.

Two other events made essentially the same point. One was the extent of state participation in the federal court appeal against FCC actions in the terminal equipment field, as previously noted. The other was a one-of-its-kind development, a four-day hearing by the NARUC Communications Committee in Washington, at which the group received oral and written testimony on the competitive scene. It then sent a ninety-three-page report to the FCC, which at the time was conducting an inquiry into the economic impact of communications competition.

NARUC said it had concluded the following:

On the basis of its investigation, including independent analyses by its staff, [it] is convinced that under current regulatory policies there will be substantial adverse impact on local exchange telephone service to subscribers resulting from the interconnection of customer-provided terminal equipment, principally private branch exchanges and key telephone systems, and from competition by the specialized common carriers in providing voice grade private line telecommunications service, and further, that the most likely competitive responses by the existing common carriers will serve primarily to exacerbate the near-term impact to the detriment of the basic exchange telephone subscriber.

Having examined the congressional outlook in the light of the convention's surcharge proposal, NARUC sent forward a revised legislative recommendation with the same objective. Its "home telephone act" would have amended the first section of the Communications Act, which calls on the FCC to "make available, so far as possible," to all people of the United States a "rapid, efficient, nationwide, worldwide wire and radio communication service with adequate facilities at reasonable charges."

NARUC's version would have directed the FCC to exercise its responsibilities in a way to "encourage, so far as possible, the establishment and maintenance of rates for telephone exchange service which are within the economic reach of every household in the U.S., for the purpose of providing the members thereof with the means of promptly summoning medical assistance and fire and police protection, and participating more fully in the business and social life of their communities."

The state regulators would have imposed on the FCC one duty which the Communications Act as written did not specify—the so-called universal service goal. Since later FCCs have taken that objective upon themselves as the justification for, among other actions, the preemption of regulatory areas traditionally occupied by the states, it could be said that NARUC obtained later de facto adoption of the "home telephone act" without Congress or any of its committees ever formally considering the proposal.

One sidelight of the events of 1974 which took place about the same time casts some light on the large number of later business failures, and abandonments of the field by other firms, seen subsequently in the terminal equipment area. The Office of Telecommunications Policy, later the National Telecommunications and Information Adminstration, asked the Donald Dittberner consulting firm to evaluate the economic significance of interconnection of customer-provided terminal equipment.

Dittberner termed many of the estimates then being made of the future growth of the customer-provided terminal industry "wildly optimistic." The firm observed: "It seems that some consulting firms believe that the users will buy interconnect systems simply to throw them out again in five to eight years—a highly improbable assumption."

By this time, the focus of activities dealing with customer-furnished equipment connected to the telephone network had long since shifted from the advisory committee to the joint board. It was agreed generally by the advisory group, however, that its development of technical criteria for the interface of equipment to the telephone network would stand as an important contribution.

Henck's comment: In all of the heated debate over whether the interface arrangements, or PCAs, were needed—an argument that always focused on the effect on customers of the tariffed charges, and the fact that PCAs were not required for telephone-company-furnished equipment even when it was identical to that available elsewhere—one thing was strangely lacking. There is no available record that anyone ever publicly discussed or questioned what was in the PCA.

It was said that the devices' principal function was to protect the system and personnel using and working on it from harmful electrical surges. This might confirm the statement of the Rochester, N.Y., Telephone Corp., proposing more liberalized tariff provisions than those advanced at the time by the Bell companies, that simple fuses would serve the same purpose.

Nonetheless, the debate ranged over flat statements that the protectors were needed to safeguard the system, a position initially accepted by the FCC on the basis of long history, to equally flat charges that the devices were simply intended to maintain the Bell companies' near-monopoly of terminals. It perhaps was an indicator of the caliber of the debate that no one, to the best of the available record, ever proposed that the makeup and functions of the PCAs be studied by technical experts to determine whether, in operation, they would in fact do what they were said to do.

Strassburg's comment: The PCAs were hastily fashioned to safeguard against harmful electrical imputs and distortive signaling from customer attachments. They were relatively simple in their design and operation, but their presence was required as the only practical buffer against harmful interconnections until they could be eliminated by responsible alternatives providing the essential safeguards.

While word was awaited from the federal courts or the combination of the federal-state joint board and the FCC, the Commission went the next mile in the Carterfone series of decisions. At issue was what appeared to be an imaginative move by a small independent company, the Mebane, N.C., Home Telephone Company, one of the many small companies which was simply listed as a "connecting carrier" in AT&T's interstate tariffs on file with the FCC. Like all connecting carriers, Mebane did not file any interstate tariff of its own, but was governed by AT&T's. It instructed AT&T to file an exception in its behalf which simply stated that it had no tariff provision governing interconnection of customer-provided terminals. No tariff provision, no interconnection, Mebane said.

In May 1975, the FCC took what was described as a "conscious extension" of the Carterfone policy in rejecting the Mebane tariff exception. It concluded that "the public interest requires that the customer's right to interconnect not be infringed merely because the device he seeks to interconnect can be defined to constitute a substitution of telephone system equipment."

The FCC declined to consider the economic harm that Mebane alleged it would suffer if its customers were free to substitute their own terminals, including PBXs and key sets, for those being provided by Mebane in accordance with the intrastate tariffs it had on file with the North Carolina Utilities Commission. The economic consequence of customer interconnection, the FCC stated, was a subject that it was still considering in the separate across-the-board economic impact inquiry in docket 20003, and Mebane was advised to make its case in that forum.

A year and a half after the North Carolina commission sought federal court review of the Telerent Leasing decision, the U.S. Court of Appeals in Richmond finally held argument. Evidently because many of the nation's large corporations, not the least of which were AT&T and IBM, had expressed their views in the case, all but one of the Fourth Circuit judges had disqualified themselves. The only Fourth Circuit judge who apparently was not a serious investor in blue chip stocks was Judge H. Emory Widener, Jr. The panel hearing the argument was composed of Judge Widener and judges from two other circuits.

A month later, at the end of October 1975, the FCC took a long step. With several major exceptions reluctantly specified by the FCC for more comment, the Commission ordered that direct connection of customer-furnished terminal equipment should be allowed. The devices would be those covered by an FCC equipment registration or self-certification program. As a condition for registration, they would have to satisfy prescribed technical standards and thus safeguard the system and its personnel.

Exceptions were main and coin telephones, private branch exchange (PBX) switchboards, and key telephone systems. The FCC said there was little technical basis for excluding them either, but allowed time for comment because their inclusion was contrary to the recommendations it had received from the joint board.

As the controversy moved into 1976, events began to move at a relatively fast pace. Essentially the same organizations that had appealed the already-argued Telerent Leasing case asked the same court to reverse the FCC on equipment registration. The joint board, as had been indicated earlier, had formally recommended that the registration program not include main and coin telephones and PBX and key systems. The recommendation's fate was clearly forecast by the fact that the four state members were joined in the 5 to 2 majority by one FCC member, Benjamin L. Hooks, but the other two FCC representatives, Chairman Richard E. Wiley and veteran Commissioner Robert E. Lee, disagreed.

It took the FCC about three weeks to reverse the joint board. This time, the vote to extend the registration/certification program to include main and coin telephones and PBX and key systems was 5 to 2 the other way.

Judge Hooks, who later left the Commission to head the National Association for the Advancement of Colored People, was joined in the minority by Commissioner James H. Quello.

In another three weeks, the federal tribunal considering the Telerent Leasing appeals issued its decision. By 2 to 1, with Judge Widener in the minority, it affirmed the FCC in its preemption of state authority over terminal equipment, concluding that the FCC "must remain free to determine what terminal equipment can safely and advantageously be interconnected with the interstate communications network and how this shall be done."

By September, almost exactly a year after the oral argument in Telerent Leasing, another Fourth Circuit panel heard the appeals to the follow-on equipment registration decision of the FCC.

Shortly before the end of the year, the U.S. Supreme Court, in a simple refusal to review without further comment, ended any question about the FCC's authority to impose nationwide standards. It let stand the appellate tribunal's decision upholding FCC preemption of state jurisdiction in the Telerent Leasing case.

Several months later, in March 1977, the appellate panel virtually ended the controversy over the scope of FCC powers and the direction in which they were being used by endorsing the Commission's decision prescribing the equipment registration/certification program. Again by 2 to 1, and again with Judge Widener dissenting, the court said, "if it is admitted—as we think it must be—that the FCC has full statutory authority to regulate joint terminal equipment to ensure the safety of the national network, then we can discover no statutory basis for the argument that FCC regulations serving other important interests of national communications policy are subject to approval by state utility commissions."

It took another six months, but the Supreme Court then officially brought the matter to a close by again briefly denying a writ of certiorari (review).

With the FCC's authority and direction now completely established, one more brief chapter remained in the telephone industry's efforts to keep some vestige of end-to-end service responsibility, as well as the economic contribution to phone companies' service embodied in the revenues from terminal equipment. This was the late-in-the-day telephone industry proposal of the "primary instrument concept." The basic local service rate, under this plan, would include at each subscriber's premises a primary or main telephone instrument furnished by the telephone company.

After the final Supreme Court ruling settled the overall controversy in the federal courts, two ranking members of the House Communications Subcommittee urged the FCC to take a close look at the primary instrument concept. At least, Chairman Lionel Van Deerlin (D., Calif.) and ranking

Republican Louis Frey, Jr., of Florida said, the proposal would permit an easier transition to the new environment. Some months later, the FCC politely started an inquiry into the subject, while warning that the plan appeared to change fundamentally the principles of the seminal decisions in the field. To no one's surprise, the Commission later concluded that even the transition had already run much of its course and that it saw no reason to reverse things and require primary instruments.

End-to-end service suffered its final coup de grace in the FCC's first two computer inquiries. The first, started in 1966, was designed to sort out and deal with the regulatory and policy questions that could be expected to arise from the growing interaction, or convergence, of computer and communication technologies.

There was a concern, shared by many, that voice-oriented facilities of the Bell System and independent telephone companies might not be responsive to data communications requirements of the new Information Age. In addition, it would be entirely possible for the telephone companies, the largest users of computer communications switching and internal information management, to sell data processing and information services to their subscribers in competition with the burgeoning data processing industry. The Commission wanted to consider whether that prospect would be, in its view, in the public interest.

Four and a half years after starting the inquiry, the FCC issued a ruling which sought to draw a line between data processing and communications services. For the first time, it prescribed structural separation to prevent telephone companies from commingling regulated communications and nonregulated data processing services. Accounting and cost allocation procedures were dismissed as too unreliable a regulatory tool to safeguard against pricing abuses that could go undetected if telephone companies could market both types of services using the same personnel and equipment.

Agreeing that many service combinations were in fact hybrid, containing elements of both communications and data processing and very difficult to classify as one or the other, the FCC opted for a relatively simple definitional test. It would ask what the primary purpose of the service offering was and apply the test on a case-by-case basis.

If the hybrid service employed data processing simply as an incident to transmitting intelligence, the offering would remain subject to regulation as communications for hire. If, on the other hand, communications was only incidental to a data processing offering, the hybrid service would be subject to strict structural separation within the telephone company.

Five years later, the Commission decided that its computer I definitions were becoming unworkable. It concluded that they were inhibiting new developments because technological change was making it more difficult to classify activities as either communications or data processing. Customer data terminals were no longer limited to input or output of information and could perform various computerlike functions as well. At the same time, these advances in technology were generating more sophisticated, or enhanced, communications services.

Thus, the FCC in computer II replaced the prior policies with a new set of classifications. All services would be either "basic" and subject to established common carrier regulation, or "enhanced," and free of such regulation. Basic services would be those offering transparent transmission capacity for the movement of information. Enhanced services would combine the basic functions with computer processing applications. All common carriers providing them—except, as usual, the Bell System—would be relieved of the maximum separation requirement upon which their offerings of similar services were conditioned under computer I.

The agency saw no reason to struggle with a system of classification where customer terminals were concerned. Instead, it decided that all customer terminals should be "unbundled" from the offering of transmission services and would not be under any kind of federal or state rate and tariff regulation. This, it was felt, would put carriers and other suppliers on a more competitive footing in taking part in the equipment market.

At first, the FCC gave the telephone companies less than two years to unbundle and detariff terminal equipment. Later, immediate deregulation, as on March 1, 1982, was limited to equipment newly furnished or sold by the carriers. "Embedded" (existing) terminals were left under regulatory control for the transitional period.

Dealing with terminal equipment and the effects of its deregulation on carrier rates, accounting, and even jurisdictional separations became nothing short of a regulatory cause célèbre for the next several years. This was particularly true during the chaotic divestiture period, when AT&T, under the terms of the consent decree, became the owner of all embedded customer terminals in the erstwhile Bell System as of the January 1, 1984, divestiture date.

On appeal, the U.S. Court of Appeals for the District of Columbia sustained the FCC rulings across the board. In doing so, it turned back the challenge of some states that the federal agency was acting beyond its authority to remove state control over the rates for terminal equipment. The court concluded that FCC preemption was justified because the objective of computer II would be frustrated by state tariffing of customer equipment.

With the court decision, one thing was clear. The institution of end-to-end service, after a century of public service, had come to a close.

12

OPEN MARKETS AND OPEN SKIES

With MCI's foot in the competitive door, it took relatively little time—as time is measured in the federal govenment—to move the door much farther open. During the first few years of the decade of the 1970s, the FCC had taken the actions that moved competitive interstate communications as far as the agency intended to go at the time.

In fact, as will be recalled in the next two chapters, competition soon thereafter went a lot further than the Commission ever intended. But that was not the expected consequence of the FCC's major moves of 1971 and 1972. At that point, the agency was still thinking of competition only in terms of private line and other specialized services tailored to the business environment. Regular long distance message telephone service was viewed as a natural monopoly by virtue of its economic and operational characteristics.

The common perception of "plain old telephone service" as something furnished by a single telephone company in each community or city, and by one integrated long distance network operated as a joint venture by those monopoly telephone firms, was so pervasive that it was simply unquestioned among regulators and communicators. If by then there were dreams, concepts, or images of competing long distance service within the MCI organization, they were never admitted, and the FCC assumed it had the authority to control competitive entry into public switched telephone service or any other service under its public interest mandate.

Given impetus by the authorization to MCI to provide private line service between St. Louis and Chicago, however, the specialized private line service bandwagon was really rolling. New applications to build microwave radio stations over routes between major cities were coming in every week. The MCI decision had sounded the starting signal for the communications version of the Oklahoma land rush. Using another Western analogy, FCC Chairman Dean Burch later suggested that some of the proposed stations would not be built and termed some applications "in the nature of mining claims."

The situation moved rapidly. In October 1969, about two months after the FCC issued the MCI authorization, the U.S. Independent Telephone Association (USITA), holding its annual convention in Washington, heard from one of its members' prospective competitors. Six years before, the state of Nebraska had set up the Nebraska Consolidated Communications Corp., or N-Triple-C, to promote and establish a noncommercial statewide system to carry educational television programs. In the wake of the MCI case, the company had gone public and sold stock to private shareowners.

N-Triple-C's legal counsel, Donn Davis, addressed the USITA convention. He forecast that "before the fifth anniversary of this talk, every state in the continental United States will have available for its use a separate statewide communications system capable of providing educational TV interconnection and general administrative and emergency communications for state government." (One thing that happened on about the fifth anniversary of his talk was that N-Triple-C was absorbed into MCI.)

In order for an alternative nationwide system to develop, he told the officials of established telephone companies, there had to be two things: sufficient demand for service and recognition by regulators. He explained, " 'Public convenience and necessity' require that competent operators be permitted to build duplicating communications systems. We believe that the second of these requirements" [the grant to MCI] has already been met. The existing toll carriers, essentially the Bell companies, will not be permitted, without competition, to absorb the exploding demand for communications services in the U.S."

An application for a microwave system is a bulky document, requiring detailed technical and topographical information on each station to be sure it is engineered to prevent harmful interference between radio signals. The applications started coming in at the FCC almost by the truckload, threatening to exhaust the Commission's filing space.

Leading the way were two organizations, one with substantial funds and a new concept, and one constantly putting together an imaginative scheme to raise a little more capital to take advantage of the entry it now possessed. They were the Wyly Corp., ahead of its time with plans for a new Data Transmission Co. specializing in data service, and Microwave Communications of America. Micom, really MCI, busily put together regional packages of proposed systems, which, like any start-up venture, targeted the busiest routes where there was the greatest promise of early profitability.

By April 1970, some eight months after the MCI decision, the FCC reported that it had on file 1,460 applications for microwave stations from specialized common carriers. These were in twenty-seven different sets, including eight regional proposals by MCI affiliates. Only a week later, the total was over 1,500, with the MCI regional system count now up to ten. The Data Transmission Co. (Datran) had announced original plans for 255 microwave stations, but later scaled down the projected system to fewer than 150, after financing became harder and harder to come by.

If the Commission were to stick with its established procedures, it was facing an administrative nightmare. As could be expected, almost all of the applications were opposed by established carriers. The usual practice was to schedule each proposed new system—or at least those which drew formal objections—for a hearing, just as the original Chicago-St. Louis applications of MCI had been. To assure that all concerned were given due process, this involved innumerable preliminary steps, extensive public hearings at which witnesses testified and were cross-examined, an initial decision by the presiding officer, eventually an oral argument before the full FCC, and—in a year or two or more—a final decision.

Even if the pending applications could be grouped in some way, the FCC was looking at perhaps a dozen major, contentious, litigious battles on the same issues, which would exhaust its resources of personnel and money. And, if things stayed about the same, and the applicants could show reasonable financial and technical qualifications, they probably would all turn out the same way the MCI case had.

While all this had been going on, another governmental development of significance was pointing the way to the future and influencing the FCC's consideration of its next moves relating to specialized common carrier policy. A year and a half before the end of his final term in office, President Lyndon B. Johnson had appointed a blue-ribbon, sub-cabinet-level task force on telecommunications policy. Headed by Under Secretary of State Eugene V. Rostow, the task force tackled a series of gut domestic and international communications policy matters and produced a nine-chapter, 450-page report in December 1968, soon after the national balloting which had made Richard M. Nixon the President-elect.

Members of the task force were, of course, all Johnson appointees, and the Nixon administration distanced itself considerably from the document. As things developed, although completion of the report and some of its contents were widely known, it was never formally issued or acknowledged by the White House. It was finally shaken loose in May 1969 by the House Commerce Communications Subcommittee, which was holding hearings on cable television and wanted to use the report in questioning Rostow.

Despite the report's unfortunate timetable on the political railroad, it was taken seriously. After all, appointment of the task force had shown recognition of a major policy need by a U.S. President, and its members were high officials from a wide swath of government agencies.

Essentially, after considerable concentration on the continuing major policy problems of the international telecommunications area, the task force drew lines around what was the accepted domestic policy framework. It first concluded that the existing integrated system of public message telephone service was sound. Beyond that, it pronounced, competition should be the rule, rather than the exception, and the regulated monopoly should be limited to functions which by their nature require control or ownership by a single entity.

A large number of private systems could raise serious problems for the integrated network, the task force conceded, but these could be met by allowing the established carriers flexibility in rate matters to meet the competition—which meant private line competition, definitely not public switched network competition. Regulatory capabilities should be strengthened to prevent destructive competition, the task force intoned, joining virtually everyone of the times in not expecting the federal courts to upset the apple cart.

Neatly generalizing, the task force concluded that prices over competitive routes should be based on costs and demand characteristics, but that new entrants should be protected against noncompensatory pricing—more popularly known today as cross-subsidy—by the integrated carrier, the AT&T long distance message telephone monopoly.

Principal dissenter in the task force was James D. O'Connell, retired Signal Corps general and the government's director of telecommunications management. Jim O'Connell reflected a consistent Defense Department line during the years of efforts aimed at breaking up the Bell System—that the nation's defense depended heavily on the integrated Bell System, which should be preserved. He viewed the task force's principal domestic recommendation—preserving the integrated public message telephone network while encouraging competition for private line service—as a "proposal to break the integrated national system into two separate parts."

This was unrealistic, O'Connell declared, "because it does not recognize the highly integrated nature of the private line and public message services, and the interactions between the two, both economically and technically. It is submitted that this recommendation, if put into effect, would be counterproductive to one of the basic general objectives identified in the task force report—more rapid innovation."

By the time the FCC issued a notice of proposed rulemaking on the subject in July 1970, there were more than 1,700 separate "special service" microwave applications on file at the agency.

Collectively, the applications presented a common question: whether the public interest goals of the Communications Act would be advanced by relatively open competition in the supply of specialized intercity services.

Under the act as interpreted by the Supreme Court some two decades earlier in the "three (international) circuits" case, *FCC v. RCA Communications*, the FCC was not free to opt for competition simply as an end in itself. The court ruled that competition was only one of several public interest factors that the FCC was obligated to consider in the total regulatory concept. Thus, it had to find that competition was reasonably feasible and could be expected to produce some public benefit in the form of more efficient and economic services.

An analysis prepared by the FCC Common Carrier Bureau concluded in favor of competitive entry. It recommended that all applicants who could meet a relatively modest set of financial and technical requirements should qualify as licensees to be given construction permits without a hearing. The FCC decided to tackle the applications as a group, by resolving the common policy issues, in a rulemaking proceeding rather than subjecting each applicant to a separate and largely redundant hearing.

In its formal notice of proposed rulemaking, the FCC asked for public comment on the questions raised by the applications and on the staff's analysis and conclusions. The proposal plowed new ground in the growth and harvesting of what has become the Commission's most commonly used procedural approach to resolving policy and other major controversial issues.

Before the specialized common carrier case, the lengthy and complex oral evidentiary hearing had been the norm in dealing with such matters. This was particularly true when an application for a construction permit or other authorization was contested, or where two or more applicants were rivals for the same authorizations, and both could not be accommodated.

The specialized common carrier proceeding changed all that. By obtaining the information it required through written comments, subject to replies by opponents and further demands for data by the Commission, the FCC could reach a reasoned and informed decision on a contested policy matter—one that would stand up in court—entirely through a "paper" proceeding. The formal trial-type evidentiary hearing became a relic of the past, except where absolutely required by the law.

In less than a year from the start of the rulemaking inquiry, in late May 1971, the FCC announced its decision, which followed the reasoning and recommendations of its staff. It would impose only minimal restrictions—the basic qualifications for any licensee—on entry into the specialized private line submarkets. It rejected arguments of opponents that evidentiary hearings on each set of applications were required, reaching the finding that it was competent to use the rulemaking approach as a matter of law, and that there was sufficient evidence in the record to support the conclusion that competition in specialized services was reasonably feasible and would result in public benefit.

How much events had moved in the short time since the closely divided
MCI decision was shown in the fact that the Commission was essentially
unanimous in the specialized carrier policy declaration. Commissioner
Robert E. Lee concurred because he had unsuccessfully proposed a three-
year moratorium on new applications not already on file when the FCC action was taken. By then, there were 1,877 station applications pending from
thirty-three organizations. Otherwise, the FCC members registered no disagreement at all with the outcome.

Public benefit would result, the FCC found, from the authorization of
new carrier entry. It concluded that this would come from new price and
service options, and that service needs would be met more rapidly. There
were, the Commission held, new or untapped markets to be served. Besides,
it found that no significant adverse effect on existing carriers' services was
likely to result.

Essentially, the nation had entered a new age of diversity in its communications requirements, spawned by technological advances. AT&T
alone could not be expected to respond with the same efficiency and timeli-
ness with which it supplied plain old telephone service.

As long as they did not cross-subsidize their competitive offerings with
the proceeds of their monopoly services, the FCC told the established carriers, they were free to compete with the new entrants. The Commission
left the specifics to future tariff proceedings, but it anticipated the cream-
skimming that was likely to occur as the new entrants exploited, at least initially, the most profitable routes. Thus, it said, "Where services may be in
direct competition, departure from uniform nationwide pricing practices may
be in order, and in such circumstances will not be opposed by the Commission."

Further, the agency addressed the crucial dependence that the new carriers would have on the facilities and services of the local Bell companies to
reach the premises of their subscribers. The FCC stressed that it expected
the local companies to provide the specialized carriers with the interconnection facilities needed so that they could render their services without
discrimination in favor of the established carriers.

The stage had been set for the major policy departure by the FCC in a
number of the developments recounted in the preceding chapters. A point
made by Strassburg in a case study of the specialized carrier case for the Urban Institute was the computer inquiry "indicated a considerable body of
public opinion that AT&T and other existing carriers had been slow in
meeting consumer needs for new and innovative communications services,
using the latest techiniques and equipment." Also encouraging the FCC to
change the existing institutional framework, he observed, was the "favorable reception" given the Carterfone decision and related rulings which
followed. "In a sense, these cases provided a current of momentum which

made it psychologically easier for the FCC to take a new initiative in the specialized carrier case.''

Everyone seemed to agree on just what it was that the FCC had authorized the new carriers to do, in both the specialized carrier and MCI decisions. The Commission and all interested parties with whom it dealt in proceedings which included a federal court appeal seemed to be talking about the same thing—those services other than message telephone and wide area telephone services. There appeared to be no need in the context of the proceeding, which was focused on the service proposals of the applicants, to define precisely just what the FCC meant by "specialized" or "private line" services. Nor was there any apparent reason to delimit just what the licenses issued to the specialized carriers permitted them to do. More to the point, as it became evident later, it was not apparent that the FCC should specify what it was that the new entrants were *not* being allowed to do and the reasons why they were not.

Prior standards of regulatory behavior were relied on in the court appeal, and once again it was demonstrated how the world had changed. There were two principal arguments put forth by the state commissions' national organization, the NARUC, joined by the Washington State Utility and Transportation Commission (Washington joined the appeal on the appellants' erroneous conclusion that the U.S. Court of Appeals on the West Coast would be more favorable to their cause). First, they said, there was no showing that service presently available from AT&T and other carriers was inadequate. Second, they argued, comparative hearings were mandatory.

The court rejected both contentions. The first, it said, gave too narrow a construction of the public interest standard of the Communications Act. It noted that the Supreme Court's 1953 decision in the "three circuits" case held that duplication of facilities is not wasteful where the competition is "reasonably feasible" and will achieve some benefit. The FCC had sufficient evidence to sustain those findings, the court stated. As to the second, it concluded that the Commission had the authority to adopt the general policy it promulgated and could do so through rulemaking.

Throughout the MCI and specialized carrier proceedings, the potential impact drawing the greatest concern was the possible effect on the established system of nationwide average pricing. It had long been accepted by regulators and other political and policy officials that the welfare of all Americans required the same rate to be charged for an interstate telephone call of equivalent distance and duration between two remote points as between two large cities connected by a "heavy" route.

Cost-based pricing over each route would obviously lead to very wide variations in per-mile rates, with customers in rural areas and small towns (those in places served by "thin" routes) paying much higher charges. From a political point of view, an obvious consideration was that thinly populated states each have two U.S. Senators, just the same as New York or California. Any threat to nationwide average pricing was bound to create a political firestorm.

Although the principal worry about possible loss of nationwide averaging by the estalished common carriers related to public switched, or message telephone, service, the obvious if sometimes tenuous relationship between charges for that service and those for private line offerings led to similar treatment for the latter. The sharply lower MCI prices for specialized private line operations over high-traffic routes, and the established carriers' clear ability to match them over the same heavy, low-cost arteries, had been frequently raised as creating a prospective difficulty.

The first directly competitive, route-priced response by an established common carrier came fairly quickly after MCI began its Chicago-St. Louis operations following the FCC authorization. It was not AT&T, but the distant number two in the private line area, Western Union, that took the step. It filed new tariffs directly matching most of MCI's medium-band rates over the route.

Warning that further departures from nationwide rate averaging might be in the offing, WU noted, "In the present tariff filing, reductions are made only in the charges for [WU's] voice frequency channel services which are threatened by MCI's offering, although it is recognized that MCI's tariff encompasses additional private line services which are comparable to various other WU services. [WU] is at present studying MCI's broader and narrower bandwidth offerings to determine the appropriate effective competitive response."

In terms of further rate filings moving away from the nationwide average approach, or in most other respects, nothing much further happened. Hardly to anyone's surprise, MCI came back with an objection charging that the rate reductions were "predatory" and that WU's reducing its rates for competitive purposes was "designed to exclude or limit competition in a market where the FCC has found that such competition is feasible and will serve the public interest." MCI, asking the FCC to reject the filing, stated, "This is not just a minor tariff change on a short segment of WU's transcontinental system, but represents the first departure from nationwide averaging of costs and rates, and could become a precedent for similar action by AT&T and the other existing carriers."

After the parties exchanged further and somewhat draconian forecasts of the effects of all of this, the FCC let the tariff changes go into effect but ordered an investigation and hearing on the subject. The hearing was never held, and the tariff revision did not have its predicted dramatic impact.

Instead, as a telegraph company tariff official of the time recalls, "We just didn't get any business. Our belief was the MCI customers were getting something in addition to the lower rates—probably a form of service similar to what later became known as Execunet—but in any event the whole controversy just sort of died out.''

The WU tariff change, still without having any noticeable effect, finally was subsumed in another controversy. AT&T, still trying to make its private line rates more competitive without shooting down the whole nationwide averaging structure, divided its voice-grade private line service points into high- and low-cost catergories, with differing rates, and submitted a new "hi-lo" tariff structure. After extensive litigation, the hi-lo proposal disappeared into yet another restructuring of private line charges. Meanwhile, WU had matched the hi-lo tariff, eliminating the specific Chicago-St. Louis rates in the process.

Shortly after the specialized carrier decision was adopted by the FCC, the Institute for the Future issued a remarkably prescient forecast of the telecommunications future. It conducted a survey of 210 industry experts, who were guaranteed anonymity, as to what they believed the U.S telecommunications industry would look like in fifteen years—in other words, in late 1986.

The experts' forecasts as to the number of telephones in service, industry revenues, and similar broad statistics were remarkable in their closeness to what turned out to be the facts. Their measure of new interstate carriers and domestic satellite systems was not all that far from the mark, and they foresaw specialized data communications networks using some facilities of the public switched network.

They missed almost totally—joining a few other distinguished members of the industry—only in two areas in which they could not read future market demand. They predicted big development of the Picturephone, suggesting that some 3 million units would be in service by late 1986. This would come, they forecast, from a major economic breakthrough in local area wideband transmission. In fact, of course, the only Picturephone market has been for business conferences, and demand has been, to put on its best face, modest.

Also, those polled, no doubt looking at the Datran system plans, saw in their crystal balls the establishment of a "new nationwide independent data network" after "several years of hearings." Datran, under the FCC's specialized carrier policy, won approval of its projected system without hearings. But the demand it visualized never really materialized, its construction financing problems were overwhelming, and only a small part of its planned nationwide network was built.

The survey by the Institute for the Future was just another of the many indicia of a firmly established line of thinking which permeated every part of the communications establishment. As did almost everyone else, the 210 surveyed experts concluded, "The regulatory climate will change, with competition expected in *some* sectors of the business.'' To those who were certain that regular long distance telephone business was such an obvious natural monopoly that no more thought need be given it, MCI was waiting with a time bomb.

In April 1972, some two and a half years after Datran's ambitious beginning, the FCC granted construction permits for a further network running from Houston to Palo Alto, Calif., with sixty-one intermediate points. Construction cost of the entire system, including microwave facilities, switching equipment, and local distribution facilities, was estimated at $280 million.

But after a series of unsuccessful efforts to obtain additional capital, not helped at all by the fact that its parent, Wyly's University Computing Co., was not finding its once-anticipated bonanza, the numbers became appreciably smaller. Datran went to the FCC in November 1973 to report a ''crucial financial juncture.'' Wyle had invested $34 million in Datran and was ''not in a position to continue to provide the funds necessary to complete the Datran system.''

The only major source of investment funds available to it, Datran told the FCC, was Walter Haefner, a Swiss national who owned a holding company named for him. He was ready to lend Datran $20 million in a rather complex financial arrangement structured to avoid the Communications Act's prohibition of alien control of U.S.-licensed radio facilities. Datra asked, and received, a declaratory ruling that the transaction was within the law.

By August 1975, Datran reported an agreement with Haefner to lend it another $10 million. The company expressed the hope that this would meet its cash needs through the end of the year, noting that both Datran and its parent were experiencing ''severe cash shortages.'' By the following June, Wyly reported ''substantial cash deficiencies and operating losses'' for both itself and Datran and announced it was looking for a merger partner and additional bank financing. At that point, Wyly said, it and Haefner had invested about $43 million and $47 million, respectively, in Datran's start-up costs.

Two months later, Datran announced that it was terminating operations as of August 26, 1976. It was put under operating receivership and followed the usual step of any communications company that went out of business during the era—it filed a $285 million treble damage antitrust suit against AT&T, charging a Bell System effort to monopolize the data transmission market.

For other carriers with system plans into which the completed part of the Datran network would fit, it then became ''bargain day.'' The receiver held an auction for substantially all Datran's assets except accounts receivable and any proceeds from the antitrust suit. What was left of Datran as a switched, occasional use, digital transmission system was ''knocked down'' to the

Southern Pacific Communications Co., now part of US Sprint, for $4.9 million. Bidding was started at $2.5 million by Western Union and moved up by increments of $100,000 until the final bid.

Once the specialized carrier policy was in place, an "open skies" domestic satellite arrangement was not too difficult to establish. As a rule, Presidents have not been noted for laying out policy options for agencies such as the FCC, particularly in memoranda which are publicly released as White House statements. But satellite communications was different. As noted in chapter 7, everyone kept getting into the act.

Thus it was that in January 1970 the Nixon administration concluded in a policy pronouncement "not binding" on the FCC but transmitted to it by the White House nonetheless, the government "should encourage and facilitate the development of commercial domestic satellite communications systems to the extent that private enterprise finds them economically and operationally feasible."

Peter Flanigan, assistant to the President, signed the memorandum. It stated, "Subject to appropriate conditions to preclude harmful inteference and anti-competitive practices, any financially qualified public or private entity, including government corporations, should be permitted to establish and operate domestic satellite facilities for its own needs; join with related entities in common-user, cooperative facilities; establish facilities for lease to prospective users; or establish facilities to be used in providing specialized carrier services on a competitive basis."

Competition continued to be the wave of the future. Within the constraints outlined in the preceeding paragraph, Flanigan said, "Common carriers should be free to establish facilities for either switched public message or specialized services." Implicit in the times, and in the FCC's later domestic satellite decision, however, was the idea that the kind of facility depended on what kind of common carrier was establishing it. The now-popular pious hope was expressed a few weeks later in the White House annual economic report. It reported "the administration's hope that increased competition will eventually make it possible to let market forces assume more of the role of detailed regulation."

With that kind of encouragement, and as satellite communications technology kept advancing, the FCC soon had another burgeoning stack of applications for domestic systems. Given the so-far successful precedent of the specialized carrier policy ruling, and the relative absence of dispute over fundamental entry issues, the Commission cranked out a domestic satellite (domsat) policy by mid-June 1972.

The Commission threw the doors to space wide open. If specified that it was providing for multiple domsat systems, rather than unrestricted open entry, but the limitations on entry were geographic and practical rather than

economic. The finite amount of space at the right altitude, and the available
orbital "slots," prevented the agency from simply authorizing all qualified
comers. (Subsequently, most applicants have been able to enter the domsat
field and operate systems, although often with different configurations and
orbital locations than they originally proposed.)

As usual, the big question related to AT&T's role and the fears, real or
manufactured, of AT&T dominance if it were turned loose as a full-fledged
market participant. AT&T's applications proposed a joint venture with
Comsat, and the limitations put on it by the FCC majority comprised one of
the major areas of dissent in a 4 to 3 vote by the commissioners in acting on
the applications.

There was widespread concern that AT&T, unlike its competitors, would
be in a position to utilize fully the capacity of a communications satellite as
soon as it became operational. Unlike any other domestic common carrier
venturing into space, AT&T could load its satellite to capacity from the inception of service by diverting its long distance telephone traffic from terrestrial circuits. This, it was felt, would give AT&T an immediate advantage
in costing and pricing specialized satellite services. In turn, this could
discourage entry by otherwise qualified groups, extend AT&T's dominance
into space, and leave in tatters the policy goal of promoting new and innovative services via satellite. Along with those worries went the perennial
alarm about cross-subsidy stemming from the combination of monopoly
telephone services and competitive offerings on the same facility.

With these concerns in mind, Strassburg urged the commissioners to consider barring AT&T entirely from the use of satellite technology, at least until others had firmly established a foothold in the market for specialized
satellite communications. A total ban, as it developed, was too drastic a
prophylactic for the Commission to impose on AT&T.

It was now three years since the MCI decision, but a sharp dividing line
continued to be maintained between the monopoly public switched telephone service and other, more specialized communications services. The
FCC therefore ruled that for three years after becoming operational, AT&T
could employ its satellites only for its monopoly services. This would give
other domsat entrants a head start which was felt by those making the decision to be needed. The FCC majority in the 4 to 3 decision concluded that
full AT&T service by satellite might "derogate from our policy of seeking to
promote an environment in which new suppliers of communications services would have a bona fide opportunity for competitive entry. This policy
was the basis for our decision in the specialized common carrier services
proceeding."

The Commission was asked to reconsider the limitation on the AT&T/Comsat
system to public switched services, and at the close of 1972, it came back

with its answer. It stuck with the head start philosophy but reminded all concerned that the ban on provision of services competitive with others would end within three years. The termination would be contingent on AT&T's divesting itself of its 29 percent interest in Comsat, which now had its own plan to launch a domsat system possibly competitive with AT&T's.

FCC Chairman Dean Burch commented that the burden would be on the other parties to show that the three-year restriction should be extended, and that extension would require a "compelling—repeat, compelling—showing."

The Commission's domsat rulings had left the way open for independent telephone companies to participate with AT&T in the planning and provision of the interstate telephone network. It authorized a system to be operated by GTE Satellite Corp. and linking the operating territories of General telephone companies. This would be the means by which those companies would be liberated from simply providing local access to and from AT&T's long lines for long distance calls originating or terminating with their subscribers.

The FCC directed AT&T and GTE to negotiate interconnection and toll settlement arrangements, but when the negotiations quickly ran into difficulties that the parties were unwilling to compromise, GTE threw in the towel. Besides, the FCC's ruling was appealed to the federal courts and was pending there. The two organizations finally came up with a plan—later approved by the Commission—under which they would join in furnishing domsat service by sharing the use of space facilities leased to them by a Comsat subsidiary. The court was asked to hold the suit in abeyance pending FCC approval of modified applications, which came in due course.

Although the Commission had specified the three-year head start for AT&T competitors in December 1972, the time did not start running until the domsat system was in actual operation. By the time the three-year deadline was due to expire—July 23, 1979—all common carrier services were subject to at least some competition.

Some of the domsat competitors did what came naturally and urged the FCC to extend the restrictions or to waive them with conditions. But, accepting the reality of the new world of competition, the FCC told AT&T and GTE that they, at long last, could furnish competitive as well as the former monopoly services by domestic satellite.

13

THE PHILADELPHIA STORY

In the specialized common carrier decision, the FCC members and staff thought they had crafted a set of new policies which were publicly beneficial, timely, and clearly defined. They were believed to provide users with diverse, competing sources of supply for private line and other specialized services. What was supposed to be a modest dose of pluralism would, at the same time, give needed credibility to the Commission in its regulation of the Bell System.

But the writers of a policy statement never think of everything. Out in the business world are all those lawyers and business school graduates looking for angles and loopholes. They have strategies and goals which may appear at first to be confounded by regulatory directives. One of their main pursuits is to pick apart documents such as the FCC's specialized carrier policy rendition and find those things which—usually because they were not specifically addressed—are effectively permitted or at least not prohibited.

From what was variously known as Microwave Communications, Inc., Micom, and MCI, there never has been a clear indication that all of what happened over the next several years was a planned step-by-step campaign. The developments that occurred were driven by an amalgam of independent, although interrelated, forces. They included what MCI did, crucial court decisions, regulatory rulings, and AT&T actions and reactions. A deliberate strategy from the outset would have had to be almost superhumanly brilliant and a near impossibility. As will be noted in the next chapter, that picture changed once some of the key underlying events took place.

All available recollections and signposts are contrary to any belief that MCI's actions were guided throughout this period by the aim of arriving as the main nationwide competitor of AT&T's 100-year-old long distance telephone service. On the contrary, MCI's founders expressly denied having any such aspirations. In their respresentations to the FCC, they discussed more modest and attainable targets in customized, specialized, or private line services. Their asserted goal and justification was to fill the marketplace voids left by the rigidities of AT&T's pricing policies and to develop new and innovative service offerings of their own.

It would seem that MCI's overriding preoccupation in the early 1970s was to raise enough money to stay in business until it could get established. It inched forward by offering any service it could to produce the always badly needed dollars, moving not by grand design but by surviving one crisis at a time.

Not surprisingly, it was MCI that first noticed the absence of any specific mention in the specialized carrier case and decision of one existing intercity service which straddled the line between the newly defined competitive specialized services and those offered by the monopoly switched network. Foreign exchange, or FX, service had been viewed by the FCC and classified and offered by AT&T for many years as one of a number of private line services.

FX service permitted a customer leasing a private line to use it to gain access to and from all regular telephone subscribers in a distant exchange. A typical example would be a New York department store with customers in Philadelphia. The store would be listed in the Philadelphia directory with a "local" number. Its patrons would make what to them was a local call at local rates to reach the New York establishment, not knowing that they were being connected to the New York premises via a Philadelphia-New York private line. FX is one of the services that permits airlines to have reservation centers covering wide areas of the nation, or all of it, while their customers in each community assume they are making local calls.

The FX line has one open end—open to all telephone subscribers in the distant exchange—and one closed end, terminating on the FX subscriber's premises. A normal private line can also be interconnected into the public switched telephone network. The switching is generally performed on the customer's premises through a PBX in which the private line terminates. This feature has caused problems for regulators, customers, and telephone companies as a "leaky PBX." In the FX configuration, the connections of the open end of the private line are made in the telephone company's central offices, rather than on the premises of the customer.

As MCI moved to position itself to go into business, it began to negotiate with AT&T for the interconnection arrangements it would need in the local exchanges to deliver its services to its customers. In some situations, it asked for local arrangements to enable it to provide its private line customers with FX services. AT&T resisted these proposals, taking the position that FX was not an authorized service for MCI under its licenses and the specialized carrier policy decision.

Neither company, however, sought an official ruling from the FCC. As early as August 1971—well before its actual inception of service and two years before the matter began coming to a head—MCI raised the FX issue at a joint meeting with AT&T, called by the FCC staff, to review the status of interconnection. The staff advised MCI to make a written request for a ruling, and assured it of a prompt response. But for reasons known only to itself and never made clear, MCI did not take any action.

Nothing more was heard by the commission from MCI on the subject for a year and a half, until February 1973. Then, one of the company's top officials, former FCC Commissioner Ken Cox, transmitted to FCC Chairman Dean Burch a copy of a letter just sent by MCI to AT&T. It set out a list of complaints about AT&T's alleged uncooperativeness in providing MCI with timely and adequate interconnection arrangements for its authorized services. Among them, it included FX, as well as a companion service, CCSA (common control switching arrangement), which AT&T had recently introduced.

CCSA was a sort of network version of the usually two-point FX. It was a network of private lines by which all users on that network could communicate with each other and also have access into the public telephone system for off-network communication. It was, for example, the foundation of the federal government's Federal Telecommunications System—a network of private lines for (mostly) official calls by government employees terminating on and off the network.

In forwarding the letter, Cox emphasized that MCI was not making a formal complaint and merely wished to keep the Commission informed of its problems while continuing to negotiate with AT&T. As a result, the FCC initiated no action on the subject at that time.

By the early fall of 1973, MCI stopped its sparring with AT&T on FX and sought an official ruling from the Commission that FX was in fact an authorized service of MCI. Its request for such a ruling came at about the same time as a related issue thrust on the FCC. The agency learned of actions being taken by AT&T which could have had unfavorable implications for carrying out the specialized carrier policy and which, if permitted to proceed, could have virtually divested the FCC of its jurisdiction over interconnection arrangements.

It has been suggested by commentators in at least two published books and argued by AT&T in its defense of Justice Department and private antitrust suits that the two issues were resolved in keeping with a scenario planned and orchestrated by Bernie Strassburg and coordinated with MCI to prop up the company's very tenuous financial status.

At this point, we depart from joint authorship for a few pages. What follows is a first-person account by Strassburg, based on his best recollection and reconstructed from relevant documents, from which the reader can draw his or her own conclusions as to the degree of validity of this characterization.

MCI lodged its written complaint with the FCC on August 27, 1973. In a letter to me, MCI Chairman Bill McGowan charged that the local Bell companies were refusing to interconnect with MCI for provision of its interstate services; that they were discriminating in favor of the AT&T Long Lines Department and Western Union as to the interconnected services they were willing to provide; and that AT&T was resorting to state commission procedures to block or delay interconnection for its interstate services. (McGowan did not specify by name FX, CCSA, or any of MCI's other contemplated services.)

I promptly sent the letter to AT&T for its response, with an emphatic reminder of the FCC's policies requiring local telephone companies to provide connecting facilities on no less favorable terms or conditions than they applied for competitive services of their own or of Western Union's. I added the admonition that where interstate services were concerned, there was no requirement to seek state approval of tariffs or contracts before providing interconnection to MCI for those services.

On September 7, AT&T replied with what amounted to a general denial of MCI's charges. It did, however, assert its intention to seek state approval for local interconnection to AT&T Long Lines services whenever it deemed state authorization to be required.

It became known soon thereafter that what AT&T did not report was that a month earlier it had already instructed the local operating companies to file tariffs with the states, rather than the FCC, to cover all facilities and services it would provide not only to MCI, but to Western Union and all other specialized common carriers, for the local distribution of their interstate services. The tariffs originally were to be filed in early September to be effective October 1, when it was anticipated that MCI should be ready to initiate its services to the public.

It was also contemplated by AT&T that the new tariffs would legally supersede its contracts covering Western Union's lease of facilities at prices which AT&T contended were substantially below meeting Bell's costs of providing those facilities. (They were in fact well below any subsequently agreed-on rates for linking the specialized common carriers.) MCI and the other specialized carriers, in their interconnection negotiations with AT&T, were demanding arrangements identical to those provided to Western Union.

In short, by filing tariffs applicable to all interconnecting arrangements with all other carriers, AT&T hoped to lawfully supersede its "exchange of facilities" contracts with Western Union and, at the same time, remove any claim of discrimination in its treatment of Western Union, on the one hand, and the remaining specialized carriers, on the other.

AT&T also made the choice to file tariffs with the states out of concern that, by filing with the FCC, it would be required to apply to the other carriers the same local link charges that were specified in its FCC private line tariffs related to interstate services of its Long Lines Department. AT&T regarded those local link charges as simply "rate elements" having no direct relationship to the cost of providing local distribution of Long Lines private line services.

The September 7 response of AT&T was followed by a fusillade of agitated oral and written complaints by AT&T's aspiring competitors to the FCC staff that AT&T was being less than forthcoming in resolving the ongoing interconnection issues.

The level of agitation was heightened on September 20, when AT&T Chairman John deButts delivered his now-famous speech to the annual convention of the NARUC in Seattle. As recounted in chapter 11, he attacked the FCC's policies seeking to foster competition in both terminal equipment and intercity services. He made it clear that AT&T regarded accommodation or conciliation with those policies as heading down a path leading to destruction of time-tested principles of regulated common carrier monopolies as the best way to meet the nation's communications requirements. AT&T, he was saying, was now determined to actively resist further implementation or expansion of the FCC's policies.

He received a rousing ovation of approval from his audience of state commissioners and regulated industry officials. Upon returning to his seat in the auditorium, he walked over to me. I knew he considered me the personification of misguided policymaking by the FCC. "No hard feelings, Bernie," he said, smilingly extending a handshake. I complimented him on the effectiveness of his delivery, while expressing reservations about the substance of his talk and whether it was in the best interests of those affected.

Clearly, John deButts had thrown down a gauntlet of defiance on behalf of the most potent corporate power in the nation. It had ominous implications for the future implementation of the FCC's pro-competitive policies and the beneficiaries of those policies.

Upon returning to Washington, I got word of AT&T's plan to replace the interconnection contracts with Western Union and the other specialized carriers with tariffs to be filed with the state commissions. There was nothing wrong with a tariff approach, and the FCC staff favored it. The Commission had already required that AT&T file tariffs with it covering terms and conditions to govern provision of terrestrial interconnection to the competing domestic satellite carriers. But the filing of such tariffs in the states would be an entirely new wrinkle in the emerging fabric of competing interstate services.

I recalled at the time a chance encounter at an NARUC meeting earlier in the year with AT&T Vice President George Cook, who was in charge of regulatory matters. He described the problems of negotiating contracts with each of the new carriers while at the same time renegotiating the agreements with Western Union. At that time, I spontaneously suggested to him the alternative approach of tariffing the interconnection arrangements. But there was nothing said in this brief exchange to suggest a possibility that state commissions administer the tariffs.

Therefore, I called on AT&T officials to meet with the FCC staff in Washington and explain the rationale for the plan if it, in fact, was AT&T's intention. On Friday, September 28, George Cook and two associates arrived with a draft letter AT&T was prepared to file that day. It would notify the Commission of AT&T's decision to, as reported, replace the contracts with state tariffs. The letter said the new tariffs would be keyed to existing exchange tariffs already on file with state authorities for intrastate private line services. It expressed the belief that this approach would achieve the objective of nondiscriminatory provision of interconnection facilities and services to all of the specialized carriers.

Using the existing state tariffs covering exchange private line services furnished directly to the public, he said, would have considerable practical value. By employing them as the base point for pricing the local distribution facilities used by specialized carriers, he argued, the across-the-board application to all users—carriers and noncarriers—would make for nondiscriminatory treatment in the telephone companies' provision of "like" services. At the same time, he readily conceded that the existing tariffs were not cost-based.

This approach immediately raised questions of legality, since the FCC had exclusive jurisdiction over interstate services. The staff would, I told the AT&T representatives, consider whether the pragmatics of George Cook's argument and the constraints of the law could somehow be reconciled, at least as an interim measure, while uniform cost-based rates were being developed for filing with the FCC.

We told the AT&T officials that the issues they had raised were of such fundamental importance that the matter would have to be reviewed promptly by the Commission. I asked some revision of the letter to eliminate any possible implication that the staff had given its blessing to the proposed course of action. The letter, as revised, was filed before the close of business the same day.

The following Wednesday, October 3, the staff had the first of two meetings with the Commission to discuss AT&T's letter and a draft reply. At the second, the following morning, the commissioners adopted the staff's draft. It totally rejected AT&T's state tariffing approach as in conflict with the law and established policy.

It noted, critically, the charges alleging refusal by AT&T to furnish exchange facilities requested by the specialized carriers for their "authorized services." It dismissed AT&T's claims that state tariffing would achieve the FCC's objectives of giving the specialized carriers interconnection on reasonable and nondiscriminatory terms. Finally, it directed AT&T to promptly file appropriate interconnection tariffs with the FCC.

AT&T, in subsequent court cases, has made much of the fact that the FCC letter did not specify what services it had in mind by its generalized reference to "authorized services." It has argued that the letter was part of a deliberate scenario, and that by simply reciting AT&T's interconnection obligations in this way, I was then free to issue my own interpretation as to the scope of those obligations and the services to which they applied.

AT&T also has made much of the fact that my staff deputy, Kelley E. Griffith, and I had several meetings with MCI in the same time frame that the state tariffing issue was before the FCC. Such meetings were held not only with MCI but also with other specialized carriers, including Western Union. They were equally at odds with AT&T on interconnection arrangements and no less alarmed than MCI by the notification they had received from AT&T that those arrangements would hereafter be governed by tariffs filed with the states.

MCI representatives, headed by Chairman Bill McGowan, visited Griffith on September 27, the day before the staff session with George Cook and the other AT&T

people. They discussed a draft MCI letter to the FCC that would, for the first time, expressly request confirmation that FX and CCSA were authorized services and, as such, entitled to Bell company interconnection.

The MCI spokesman also discussed with Griffith the possibility of going directly to the federal courts as a means of getting a faster interconnection order than they might expect from FCC processes. MCI was hoping to go into broad-scale commercial operation within the next couple of months. It was assumed that a letter from the FCC confirming their rights to FX/CCSA interconnection might support their case for judicial relief.

Kelley Griffith reported the meeting to me, and we agreed that FX and CCSA were specialized services clearly within the scope of FCC policies and the MCI authorizations. A significant factor in this conclusion was that FX/CCSA accessed the public switched telephone network in the same way that private line services were equally capable of doing. The only difference was that the former took place in the telephone companies' central offices, and the latter occurred on the customers' premises. Moreover, FX, for many years, had been classified by AT&T's own tariffs as a private line offering.

We felt, however, that the more pressing and fundamental issue was jurisdiction over the Bell companies' interconnection tariffs. We thought MCI would be better advised to defer any request for a bureau ruling until the jurisdictional question was past. MCI's raising the FX/CCSA dispute at that moment, we concluded, could lead to complications and delay. Griffith so advised McGowan, who left the FCC without filing his letter.

Following the first of the FCC's meetings on the state tariffing question, McGowan and Senior Vice President Ken Cox paid me a brief visit. I told them of the bureau's position and of the draft letter to AT&T, which, if adopted by the Commission, would affirm their position. This was the letter the FCC did in fact approve the following morning.

Also, I indicated that when and if the Commission approved the draft letter to AT&T, I would be prepared to confirm in writing that FX and CCSA were within the scope of the FCC's policies and MCI's authorizations. MCI could then pursue whatever judicial or administrative remedies it thought most effective to give it the needed facilities to initiate operations as a specialized common carrier.

In so strategizing at this time, I was constrained by a mix of considerations. They went to the predictability of the Commission in holding firm to its pro-competitive policies and to MCI's business need for speedy relief. The commissioners, individually and collectively, were being increasingly buffeted by the telephone industry and state commissions. The latter were becoming more outspoken in their attacks on the FCC's policies of fostering competition in both specialized intercity services and customer premises equipment.

John deButts's speech still reverberated and underscored the difficulties that lay ahead for carrying out the FCC policies. In the wake of the intercity/state offensive, I sensed a growing uncertainty among some of the commissioners about the efficacy of competition. They were troubled that the new carriers had not made more dramatic progress in introducing new and innovative services, let along competing for existing markets.

The commissioners were creatures of politics—some more than others—and there were those who believed AT&T's goodwill, or the lack of it, could make a difference in realizing their political fortunes. It was not unknown for a commissioner to shade his or her voting, even if it meant the compromise of a personal conviction, to avoid incurring AT&T's antipathy.

With this amalgam of forces—whether all real or partly illusory, I will never know—I was concerned for the viability of the competitive policies I had been instrumental in shaping. I therefore felt justified in providing the fledgling carriers with all the help I could extend within the limits of my authority and powers as bureau chief. The authority to interpret and implement FCC policies was, of course, my most important and useful asset.

As we return to joint authorship, it can be said under any format that events moved rapidly after the FCC approved the draft letter telling AT&T it would have to file its interconnection tariffs with the federal agency. AT&T submitted the tariffs under protest, offering about the same rates that had been contained in the state filings for intrastate services.

(It was more than two years later, as the culmination of months of bitter wrangling in negotiating sessions between the parties, that AT&T and the OCCs (other common carriers, as they were now known) hammered out a compromise agreement under FCC aegis on the charges to be made for the local facilities. What was supposed to be an interim agreement until cost-based tariffs could be prepared lasted for almost a decade.)

Almost immediately after AT&T filed the tariffs in response to the FCC's directive, MCI showed up with its previously discussed request, asking the FCC to confirm that the connections it could obtain included those for four enumerated services, including FX and CCSA. Within four days, on October 19, Strassburg replied in the affirmative.

Strassburg has said in testimony prepared for an antitrust case (incidentally, as a prospective witness called by AT&T to support its position that MCI's damage award demands in the second Chicago trial were excessive) that he "might have handled the dispute differently" under other circumstances. "Specifically, I may well have referred the entire matter to the Commission itself instead of responding directly to MCI as bureau Chief." In addition to the "growing uncertainty" mentioned in his comments here, he "thought time was of the essence" because MCI was about to begin service beyond the Chicago-St. Louis route.

Both MCI and AT&T lost no time in reacting to Strassburg's interpretation that FX and CCSA were authorized MCI services with which the Bell companies were obliged to interconnect. AT&T asked the FCC to review the ruling, as well as the Commission's earlier directive of October 4 that interconnection facilities were to be covered by tariffs filed with the FCC and not with the states. At the same time, Western Union asked the agency to prevent AT&T from using tariffs to abrogate its exchange of facilities contracts with Western Union.

MCI, without waiting for any further developments at the FCC, but with Strassburg's favorable letter of interpretation in hand, was off and running to a U.S. District Court in Philadelphia, asking the court to issue a mandate requiring the Bell companies to provide them with the interconnections needed for their service offerings. MCI apparently chose Philadelphia as its battleground because its potential customers planning to use FX and CCSA were heavily concentrated in Pennsylvania.

Within a few weeks following AT&T's protest and Western Union's plea for protection of its contract with AT&T, the FCC moved on its own. Noting that all of the matters before it raised questions about the lawfulness of actions taken by the Bell companies related to interconnection facilities, the FCC directed AT&T to show cause why it should not be ordered to cease and desist from the conduct complained of.

Both in court and before the FCC, AT&T did not rest entirely on the contention that FX and CCSA were public switched or regular telephone services because of their connection into the general telephone network. In addition, it argued, MCI was seeking to become a telephone company, a role for which it was not authorized, providing "joint through service"—a longstanding regulatory term when two or more telephone companies joined together—in a joint offering of service to the public. AT&T was emphatic that none of its companies had any such intention where MCI was concerned.

AT&T told the FCC that the types of local interconnection sought by MCI

are somewhat obscure, but seem to include connection by MCI into the message telephone service and the insertion of MCI facilities into telephone company CCSA networks. . . . Such requests indicate that MCI, far from providing "specialized" services of its own to the public in competition with the established carriers, simply seeks to operate, at least insofar as foreign exchange and CCSA services are involved, as a connecting carrier in providing joint through service with the Bell companies. . . . The Commission's [specialized carrier policy decision] makes it abundantly clear that MCI is authorized to provide specialized services in competition with, and not as a joint participant with, the Bell companies.

The outcome of the two-day hearing in Philadelphia, and subsequent appellate proceedings, was later to bulk large in antitrust suits, both private and those initiated by the government.

In a two-day hearing in the MCI court case before Judge Clarence C. Newcomer, both sides presented witnesses supporting their previously established positions. Among those called was Kelley Griffith of the FCC, who reviewed events at the Commission as recounted earlier in this chapter. Because all concerned stuck to their guns, the testimony may be most noteworthy in considering MCI's present status as a multi-billion-dollar corporation. Bill McGowan's testimony included MCI's balance sheet at the time—close to the end of 1973. It showed $7.5 million in current assets and $10.6 million in current liabilities. Monthly expenses were $2.4 million. Revenues were $48,000 a month, but it was noted that orders pending or being processed, if the requested connections were made, would add $800,000 to that figure.

Soon after the start of 1974, on January 7, Judge Newcomer granted the full injunction sought by MCI. He ordered the Bell companies to provide MCI with connections to their local networks for MCI's interstate foreign exchange service, local transiting facilities between specialized common carriers on an interstate through basis, "interstate private line services connecting common control switching arrangement facilities . . . as is now being done for AT&T when its Long Lines Department offers this service," and interstate services to cities near MCI terminals, but outside the Bell System companies' local service areas.

The next series of court manuevers cost AT&T a large part of the subsequent MCI antitrust judgment and clearly contributed to the outcome of the Justice antitrust complaint heard by Judge Harold H. Greene in Washington. It contributed to the charge that AT&T used its control of the commonly owned "bottleneck" local companies to engage in unfair and predatory competition in interexchange service, in which AT&T, the owner of the local bottlenecks, also was one of the competing interexchange carriers.

AT&T sought a stay of Judge Newcomer's injunction, under which it had made the connections sought by MCI. On February 4, the chief judge of the U.S. Court of Appeals for the Third Circuit, headquartered in Philadelphia, granted a temporary stay of the injunction, until a panel of the court had an opportunity to consider the request.

When the injunction was stayed, the Bell companies "pulled the plug" on nine MCI customers, after giving them time to make other service arrangements. MCI later convinced the courts that this action damaged its reputation as a stable, permanent provider of service.

Two weeks later, the appellate court panel denied AT&T's requested preliminary stay by a 2 to 1 vote, but its action was in turn stayed while the full court took up a petition for reconsideration. Later, the court denied reconsideration, but went ahead with its consideration of the premanent stay being asked by AT&T.

By mid-April, the appellate court issued its order, in essence putting the whole issue back into the FCC's lap. A three-judge panel concluded that the Commission had not really decided the dispute and noted that the FCC letter of the prior October 4 did not specify FX and CCSA. Furthermore, the judges ruled, the specialized common carrier decision showed, in contrast to what was being argued, "no significant claim by MCI of the right to have

FX, CCSA, or similar interconnection devices, as opposed to the term 'local distribution of proposed services.' " The court noted that the FCC letter to AT&T "was primarily focused on the contention of AT&T that tariffs covering interconnection with MCI should be filed with the state commissions, as opposed to with the FCC."

A week later, the FCC was back with its decision in its show cause proceeding against AT&T. It volunteered that there may have been room for doubt about the application of its prior directive, although its decision gave no indication as to the basis for that doubt. But this time it left no doubt as to the pervasiveness of its pro-competitive policy in specialized and private line markets, and it firmly directed the Bell companies to provide distribution facilities to the OCCs (other common carriers) similar to those furnished Long Lines, within ten days of the issuance of the order.

Within another week, early in May 1974, the Court of Appeals in Philadelphia briefly denied AT&T's request for a stay of the latest FCC directive. The Commission order became effective May 3, and AT&T filed the new tariffs as ordered. AT&T also reported that it would promptly begin reconnecting the MCI customers whose service had been disconnected when Judge Newcomer's order was put under temporary stay.

The FX/CCSA controversy ended in mid-September 1974, when the Third Circuit affirmed in full the Commission's order of five months earlier requiring the Bell companies to furnish the OCCs nondiscriminatory interconnection. The dispute over what it all meant, and whether the parties' actions had been properly based, went on in antitrust courtrooms for some years.

14

BY ANY OTHER NAME

MCI realized, at some point before or during the foreign exchange/common control switching arrangement episode, that its potential for growth and profitability would be seriously limited it if continued to be confined to the specialized markets. At what point the realization came is not known, and probably never will be unless Bill McGowan's memoirs are published.

Two things are, however, perfectly clear. First, MCI's total success in the use of government processes, both administrative and judicial, to win the FX/CCSA battle was achieved without tipping its hand in any way to whatever goal it had to enter the public switched telephone market. Second, it then quickly embarked on a plan of testing the FCC pro-competitive policies to their outer limits.

Experience has demonstrated that no specialized carrier could support a nationwide network of its own facilities without depending principally on the big money and steady growth inherent in conventional long distance service. Survivability of carriers such as MCI depended on tapping that lucrative market served exclusively up to them by AT&T and its independent telephone company partners.

The question was whether the specialized carriers would be allowed to do so, in light of the strictures generally taken for granted as having been imposed in the FCC's MCI and specialized carrier policy decisions.

By winning the FX/CCSA battle, MCI had established a beachhead from which to launch an all-out assault for the huge prize of long distance service. This time it would not, it was evident, have the FCC on its side. On the contrary, the Commission could be expected—as it did in fact—to take the stand that MCI was trying to invade a territory clearly beyond the bounds of the established FCC policies and the terms of MCI's authorizations.

Regardless of the extent of detail in which its scenario might have been written at the time, MCI wasted little time in making its moves. The tariff under which its first competing long distance service, known as Execunet, was offered was filed with the FCC within weeks after the U.S. Court of Appeals for the Third Circuit affirmed the Commission order that the Bell telephone companies had to furnish the OCCs (other common carriers) with the same interconnection arrangements they provided to the AT&T Long Lines Department.

Whatever MCI's intentions were, they were not at all obvious. The company masked and screened its strategy with consummate skill. As Strassburg has said in his "offer of proof" mentioned in the preceding chapter,

In both the MCI proceeding and the specialized common carriers proceeding, MCI had explicitly represented to the Commission that it was not seeking authority to compete with AT&T's messege toll service and wide area telephone service—AT&T's regular switched voice long distance services. Based on these representations of the limited nature of MCI's proposed service offerings, there was no need for the Commission in those proceedings to address the multiplicity of difficult economic and public policy questions that competition with MTS and WATS so obviously raised. . . .

In the event that MCI had proposed Execunet service to the FCC while I was bureau chief—as part of the FX/CCSA interconnection dispute or independently of that dispute—I would have advised MCI that its facility authorizations and FCC policies precluded such a service. If MCI chose to pursue the matter further, I would have recommended that the Commission take appropriate action to prevent MCI from providing Execunet service on the ground that it was not within the family of services authorized by the FCC to be provided by the specialized common carriers, and that therefore MCI was not entitled to use FX/CCSA interconnections to provide Execunet service.

MCI's concept of Execunet was so strikingly simple that it is surprising that the FCC failed to anticipate it and condition MCI's authorizations to prevent its happening. Remember that FX is open at one end of a two-point private line circuit, giving the customer the ability to have communication with all telephones in the exchange at the open end.

So what is Execunet? Open both ends of an MCI private line circuit so that MCI customers in each city can dial up, via the MCI switches, any telephone in the other city. Treat the private line as if it were being used in common by all customers, and you have MCI's version of a shared, FX private line service. This is what MCI named Execunet.

Now, all telephones in the exchanges on both ends of the shared circuit are able to reach each other, which of course is the hallmark of the public switched telephone network. To users, the main difference is that they have to feed up to twenty-four digits into their stations, compared with ten if they used the Bell System network entirely.

That a government agency in its policy planning cannot anticipate every contingency, particularly the responses of the courts, was demonstrated dramatically in the Execunet controversy. Most graphically, that demonstration came in the remarkable court decision which must be rated as an extremely significant event in restructuring the communications industry—second only in importance to divestiture itself.

It was all the more remarkable because policymaking in our system of government is intended to be the province of Congress and administrative agencies carrying out the laws passed by Congress. The judiciary is to adjudicate disputes (in appellate review of administrative agency decisions, the question of whether the agency acted in accordance with the law). The Execunet case is a classic example of a court's exercising its powers to install its own policies in place of those clearly intended by the administrative body.

As we observed, regulatory agencies are not omniscient. A good example was the FCC's handling of the MCI tariff which created Execunet and which was filed soon after the Court of Appeals ended the FX/CCSA controversy. It was an unusual tariff described by MCI as "modular." It would give the customer what amounted to a menu. The bill of fare would list all of MCI's authorized services. The diner would make his selection from the list of appetizers, entrees, salads, and desserts. Who could disagree with an a la carte approach to telecommunications services, since everything the customer could choose was appropriately authorized?

FCC tariff examiners had never seen anything like the modular tariff before. They did not quite understand it. At the same time, they really did not see any basis to object to it. MCI's competitors also could not find any basis for objection, when a protest might have put them in the position of complaining about the communications tariff equivalent of baseball, Chevrolets, and mom's apple pie. The modular tariff, essentially unquestioned, routinely went into effect.

By the spring of 1975, MCI was providing Execunet service to some of its large customers, who were happily enjoying lower rates for regular long distance calls, over heavy-traffic, low-cost, high-volume routes, in their first experience with competitive long distance service. All they needed, besides an MCI account number, was a TouchTone, or touch-signaling telephone, not one of those with the conventional dial wheel. In fact, a dial wheel telephone could be employed to do the necessary signaling, if one covered the dial with a suitable touch-signaling conversion pad.

It did not take AT&T long to find out what was happening, and it took even less time for stunned members of the company's federal regulatory department to rush to the FCC with the news. Bearing an MCI account number and a touch-signaling pad, they quickly alerted FCC officials by putting on demonstrations of the ease with which the Chicago weather report number could be called, using something which looked like, acted like, and in fact was like regular long distance telephone service.

Urged to file a formal complaint, AT&T Vice President James R. Billingsley took little time. He wrote to the Commission pointing out that MCI "is currently offering long distance message telephone service, which it is not authorized to provide and has not tariffed." Jim Billingsley, barely able to conceal his indignation, told the FCC that "MCI's Execunet service is simply long distance message telephone service. MCI had never sought authority to enter the long distance telephone business—in fact, it has said it did not intend to provide long distance toll service." The FCC Common Carrier Bureau asked MCI to reply within fifteen days.

Execunet closely resembles any foreign exchange service, MCI Vice President and later President Bert C. Roberts, Jr., said in reply to the Commission. He reminded the FCC that the tariff components became effective the preceding October without any protest.

He agreed that Execunet "can be used as a limited alternative to long distance message telephone service." But, he went on, "The same thing is true of any foreign exchange service—and, indeed, of any private line service—and the results are exactly the same whether MCI or AT&T provides this alternative mode of communications. The only way long distance message telephone can be sheltered from the effects of its cross-elasticity with other communications services is to prohibit all such alternatives and force everyone to use Bell's DDD [direct distance dialing] network as the exclusive means of intercity service."

Furthermore, MCI was still shying away from any suggestion of full-blown long distance competition. Execunet, the FCC was told, "is a service for business and other similar entities having significant communications requirements between a relatively few communities. . . . Execunet is a specialized service customized to fit the needs of business and other similar users."

Commission action on the AT&T complaint came in slightly less than a month. In early July 1975, the FCC ruled that Execunet was essentially switched public message telephone service and must be eliminated from the MCI tariffs on thirty days' notice. The agency said in a letter to MCI, "Since your authorizations are limited to private line services, you cannot lawfully tariff and operate other services on these facilities. As a result, your tariff no. 1 is hereby rejected insofar as it purports to offer Execunet service, but without prejudice to MCI's offering any other service which you are authorized to provide."

MCI promptly went to the U.S. Court of Appeals in Washington, asking that the tribunal stay the effect of the FCC order while it considered the merits of the dispute in a full-scale review. Almost immediately, a two-judge panel of the court issued a temporary stay order, giving Execunet at least a reprieve. One of the judges taking the action was J. Skelly Wright, whose name is now inseparably linked to the new telecommunications industry structure.

Several months later, the court granted an FCC motion to hold its review in abeyance. The Commission asked that the case be sent back to it so that it could consider "new and more elaborate" arguments raised by MCI.

The FCC took its time considering those new arguments, but the concept that public telephone service was a natural monopoly, as old as the twentieth century, was not one that could be overturned easily. An oral argument was held in late May 1976, after which the commissioners issued instructions to the staff to prepare an order reaffirming the one issued the previous July. Execunet, the FCC said in a brief statement, was "not within the scope of the present authorizations granted to MCI."

Representing MCI at the argument, former Commissioner Ken Cox commented that the AT&T message toll service MCI was charged with duplicating is "inseparable with an enormous grid of facilities," and that "nobody in his right mind" would try to duplicate the AT&T public switched network. Execunet, he argued unsuccessfully to the FCC, was merely shared private line service.

Once again, the Court of Appeals put the FCC order under stay. AT&T tried to get the stay lifted, and MCI said in reply that such an action "would put it out of business." MCI reported that it "has not yet attained positive cash flow. It would not be able to do so if the court's stay were dissolved or modified."

In October, the stay was in fact modified, but in a way which permitted MCI to keep its present Execunet customers. AT&T then protested that MCI was using the stay to build up business in the challenged area, when the court intended the stay merely to maintain the status quo during litigation. The court agreed on that point, "grandfathering" the existing customers for the time being, and ruling that the FCC order was in effect as far as banning any further soliciting, marketing, or distributing of Execunet service was concerned.

Meanwhile, observers watching closely for signals of events to come thought they saw one at the Court of Appeals. As it developed, those who had that opinion were wrong. At issue in another case was the FCC authorization to the ITT Corp.'s fledgling intercity carrier, United States Transmission Systems. A three-judge panel issued an opinion written by Judge Wright affirming the FCC action and stating that the FCC policy was "full competition over the entire range of private line services." Further, it was observed, the Commission contemplated that the specialized carriers "would offer new and different types of service which would expand the total communications market and not significantly divert traffic from existing markets."

It thus appeared, in a classic instance of turned-around signposts, that the FCC and the Court of Appeals had the same perception of the agency's policies.

Nearly a year later, on July 28, 1977, and after the customary legal briefs and oral argument, two other judges of the Court of Appeals joined in a unanimous opinion written by Skelly Wright. Their conclusion might be summarized as one holding that since the FCC had not specifically prohibited MCI from providing Execunet, the company was therefore allowed to furnish the service.

The court declared that MCI's facility authorizations "are not restricted, and therefore its tariff applications could not properly be rejected." It was noted that the FCC did not issue any general service restrictions in its specialized carrier policy order, or include any in the licenses granted to the carriers, "although this would certainly seem to be the natural thing to have done had the Commission sought to restrict specialized carriers to specialized private line service offerings."

Whether the specialized carriers were authorized to provide services other than "those immediately at hand"—including, of course, Execunet and others like it competing with switched long distance service—must be decided by the FCC in other proceedings, the judges declared. The Commission's decision rejecting the Execunet portions of the MCI tariff was reversed and remanded to the FCC.

The court looked for specific findings which, of course, simply were not there because the FCC had always taken them for granted. The judges said, "Because the FCC had not so far determined that the public interest would be served by creating an AT&T monopoly in the interstate [message toll telephone] field, it may not properly draw any inferences about the public interest from the bare fact that another carrier's proposed services would compete in that field."

When the FCC granted the specialized carriers' authorizations on the basis of a finding that such an action was in the public interest, Wright went on, "The Commission did not perhaps intend to open the field of common carrier communications generally, but its constant stress on the fact that specialized carriers would provide new, innovative, and hitherto unheard-of communications services clearly indicates that it had no very clear idea of precisely how far or to what services the field should be opened."

Further, the court commented, the staff report which led to the specialized carrier policy decision indicated that "a decision was apparently made to consider the consequences of future developments in appropriate future proceedings."

The foundations of a lifetime of conventional wisdom crumbled as the FCC and AT&T forlornly played out their hands, seeking at least to control the damage. Locking the barn door, the Commission began a policy of not issuing any further permits to specialized carriers unless they provided specifically and solely for private line service.

The process of seeking Supreme Court review was launched, and it would later be noteworthy that during this entire period of lower-court decisions, which shattered the fabric of the nation's telecommunications structure, the high court consistently refused to hear any appeals. As the Execunet controversy went on, the Supreme Court twice turned down pleas for writs of certiorari (review) without comment.

An insight into the FCC's state of mind was provided in the Commission's brief filed with the Supreme Court at this time. It declared, "MCI, in effect, is telling the FCC that it might have been tricked into opening ordinary long distance service to competition, even though MCI and the other applicants for entry had solemnly assured the Commission and the courts that the only services in question were private line services."

In January 1978, about the same time that the Supreme Court turned down the review request without comment, AT&T declared that it would provide no more connecting facilities to OCCs which could be used for regular long distance service. The company asked the Commission to issue a declaratory ruling (statement of interpretation of law or policy) that the Bell companies had no obligation to provide local connections not used for private line service.

Toward the end of February, the Commission, sticking by its guns up to now in the face of the court remand, issued the declaratory ruling AT&T had asked. It also announced it would begin an inquiry into the question the Execunet court said it had not addressed: whether the public interest required the switched public message services, message toll and WATS, to be furnished under monopoly or competitive conditions.

MCI hustled back into court with the declaratory ruling, and the appeals court fairly promptly, by mid-April, reversed the Commission's action again. It approved an MCI motion directing the FCC to comply with the court mandate and require interconnections for Execunet service.

In about the clearest statement yet of its ruling that Execunet was allowed because it had not been banned, the tribunal stated, "MCI facilities authorizations encompassed Execunet service precisely because the specialized carrier policy decision did not explicitly and affirmatively exclude this type of service from consideration."

Once again, the Supreme Court, after refusing to stay the appeals court's verdict, routinely refused to review the decision without comment. Long before, in June 1978, the FCC had returned to regular processing of the specialized carriers' construction permits, issuing what now, by court ruling and the law of the land, were wide open authorizations for all types of communications services.

Technically, the newly issued permits were subject to the outcome of the market structure inquiry in which the FCC supposedly was going to consider whether "plain old telephone service" should be a monopoly or a competitive undertaking. But the world had moved on, and the issue, for all practical purposes, had been decided.

Changes in leadership of the FCC and its staff had brought changes in philosophy regarding the role of government in the marketplace. In communications, as in air and land transportation and other areas of economic activity that had been controlled by government, regulation was now becoming regarded as a wart on the face of progress and prosperity. The competitive marketplace was increasingly viewed and cited as the optimum solution to matching supply and demand in communications equipment and services as well as for any other issue or problem which arose.

Buttressing the new attitude toward regulation was, of course, the pending government antitrust suit to break up an already eroding AT&T monopoly. It was, however, to be a few years before the trial of that lawsuit significantly picked up in tempo.

15

FROM ENFIA TO ACCESS

Throughout its history, the FCC has jealously protected its jurisdiction and sought to maximize use of its power in pursuit of its regulatory goals. With rare exceptions, the appellate courts have deferred to the agency's construction of its authority and have not often reversed the Commission's decisions.

Thus, it was not easy for the Commission to accept what rapidly was becoming a fact: that someone else—the Execunet court—had authorized competing carriers to serve the long distance telephone market. What was more, they not only had been authorized to do it, they were in fact doing it. Company after company, mostly licensed during the specialized, common carrier "Oklahoma land rush," announced entry into competitive telephone services at rates significantly lower than AT&T's.

Grimly, the FCC maintained its position that it was investigating the economics of competition in one inquiry, and whether it was in the public interest to have competitive long distance telephone service in another. The market structure inquiry into the relative public interest desirability of monopoly or competitive message telephone service (MTS) and wide area telephone service (WATS) was docketed number 78-72. For reasons not related to its original and supposedly fundamental purpose, the docket number was to become one of the best known in FCC history.

Because it simply had not reached that primary question, and the court had remanded the issue to it, the FCC said with a straight face, all of those wide open facilities authorizations now being issued were subject to the outcome of docket 78-72. The agency's position was that it had the authority and responsibility to decide whether interstate telephone service was a monopoly or a competitive service, and that if it decided in favor of monopoly, all of the burgeoning competitive businesses would be dismantled, and the competitors would either return to specialized private line services or resale operations, or else go out of business.

However much the die was already cast, and long distance telephone service appeared to be irrevocably in the competitive mode, the notice of inquiry in docket 78-72 raised some relevant and pertinent issues which are still seeking answers. It is the effort to find those answers and then fine tune them in light of changing conditions that causes the 78-72 designation to be such a long-running phenomenon.

When the FCC issued its notice of inquiry in early 1978, it fairly well spelled out what would be the key questions in a competitive long distance environment. What has proved to be the most contentious is the reimbursement the interstate service should make to the local operating companies for the use of local plant in originating and completing long distance calls.

For most of its history, long distance telephone service was an integrated, unified undertaking of the Bell System, working in conjunction with a large number of independently owned local companies. The interstate part of the service was governed by a single nationwide schedule of rates, designed by AT&T subject to FCC oversight, intended to produce the revenues to compensate each of the companies for its costs of participation, and providing a uniform rate of return for each.

Thus, a local company's share of the interstate pool of revenues depended on the allocation of costs between local and long distance services. As noted earlier, the allocation process was controlled by the jurisdictional separations process—a chronic source of controversy between federal and state regulators. The more local costs were recovered from the interstate pool of revenues, the less was left to come from local ratepayers, and exchange rates would be lower.

Obviously, the new, intensely competing long distance carriers were not partners of the local companies. MCI, Southern Pacific Communications (Sprint), ITT's U.S. Transmission System, Western Union's revived but sputtering telephone service, and the others were strictly customers of the exchange companies, using the local service to reach their own subscribers. Like all customers, they wanted maximum service at minimum prices.

Closely related was a second question posed by the FCC: What additional charges, if any, should be levied on interstate services to support local exchange operations. It was taken for granted by everyone except some of the new competing long distance carriers—who had very good financial reasons to contest the common conception—that long distance revenues lent some measure of support or subsidy to local service. Without that contribution, it was agreed by all but the competitors, local rates would be considerably higher.

Another key issue was equally sensitive and, to legislators and policy-makers with rural or small town constituencies, perhaps more sensitive.

This was the essentiality and public interest significance of maintaining uniform rate averaging in the interstate toll schedules. Its importance in keeping the same per-mile rate for a call between two farming communities over a high-cost, low-volume route as between two big cities via a low-cost, heavy-traffic artery could hardly be overstated to the members of Congress from low population areas and to public service commissioners in any state with a significant rural or small town population.

When long distance telephone service was a monopoly, nationwide rate averaging, although never officially approved by the FCC, was taken for granted. But the institution was due to come under increasing pressure as the competing carriers planned their systems to link heavy traffic centers. They were under no political obligation to serve small communities and rural areas, and they gave no sign that they intended to do so. (Ironically, it is the AT&T network which permits MCI, US Sprint, and the other competitors to complete calls for their subscribers anywhere in the nation. When they get off their main routes, they usually obtain AT&T service and resell it to their customers.)

The principal remaining issue would have been relevant only if the FCC had decided that long distance service should be a monopoly, and if it had been permitted to enforce that decision. It was the extent to which MTS (regular long distance) services and facilities might be severable from other carrier operations, and whether it was possible to provide WATS competitively while MTS was a single, integrated offering.

But while the Commission held doggedly to the belief—in public, at least— that the monopoly versus competition question was still one for it to decide, a host of converging forces kept the heat on in favor of a competitors' fait accompli.

The political tide was clearly turning against AT&T and in favor of its competitors. As will be seen in chapter 17, the question on Capitol Hill was becoming not one of whether any legislation would favor or disfavor AT&T, but rather whether AT&T and its allies could stave off an unfavorable bill giving the advantages to its competitors.

Economists, who controlled government theory during this period and were successful in advancing textbook arguments without much in the way of real life demonstrations, were in the saddle and pushing hard for deregulation. Out in front was Alfred E. Kahn, who had become chairman of the Civil Aeronautics Board just in time to dismantle it while the airlines were deregulated. Fred Kahn was not only a prominent economist—among many other associations he was a member of AT&T's council of economic advisors—but was well versed in government maneuvering as chairman of the New York Public Service Commission, CAB chairman, and the Carter administration's chief price stabilizer.

Deregulation, of course, meant competitive provision of utility-type services. As a governmental surrogate for market control by competition, regulation went hand in hand with monopoly provision of services.

We now come to one of the very few points in this lengthy recounting of fifty-three years of history where we as authors have an irreconcilable difference. It relates to the situation faced by the FCC after the Execunet decisions of the court. Technically, as noted in the preceding chapter, the Court of Appeals had not ruled in favor of competitive long distance service and had simply stated that the FCC had made no findings in favor of monopoly.

> *Strassburg's comment:* The FCC, rather than sparring with the court, could have come to grips with the gut public interest issues presented by competition in long distance telephone service. It would not have been too difficult to confirm its long-standing conviction and to make timely findings that could not be overturned by the court that such competition would run counter to the public interest in the maintenance of uniform nationwide rates, integrated planning and operation of a unified nationwide network, and separations principles and revenue settlements that supported affordable exchange service. While these questions were being examined, the FCC would have had no problem in freezing its authorization of any new facilities to the extent they could be used for long distance telephone service. A decade of actual experience, as I discuss more fully in chapter 21, tends to substantiate the validity of these concerns.

> *Henck's comment:* The thrust, tenor, and approach of the court's rulings was such that it would have come close to a belief in the tooth fairy to expect court affirmation of any subsequent Commission ruling that long distance telephone service should return to a monopoly status.
>
> There was little reasonable ground for the belief that the FCC, if in its wisdom it concluded that long distance telephone service should go back to being a monopoly, historically conducted by AT&T in the public interest, could make the decision stick. The chances of a congressional enactment to the contrary were excellent; if for some reason that did not occur, court reversal appeared highly likely.

Nonetheless, while the next act of the drama was awaited, there were practical problems to surmount. First in the minds of the still-integrated Bell System companies was the need for getting some compensation from the competing long distance carriers for the use of their local property and services.

By mid-1978, the heated struggle between OCCs (other common carriers) and the Bell System over the amounts the OCCs should pay for their use of the pervasive national telephone network was dominating the communications scene. AT&T was still trying, generally unsuccessfully, to find the key to the cost-based tariffs which were supposed to replace the compromise interim rates for facilities employed for OCCs' private line service, negotiated in late 1975. The difficulties this project faced were exemplified by a later string of court cases on the subject, which continued for more than a decade.

The difficulties were threefold. First, any rates calculated on the basis of accounting studies of a complex, multi-purpose system were bound to be the subject of hot controversy.

Second was the low base the two sides were starting from. Historically, the only such rates in existence were those incorporated in the exchange of facilities contracts between the Bell companies and Western Union. A lot of inflation had occurred in the long years since those agreements were reached. And they were not necessarily the products of extremely hard bargaining. Western Union's always straitened circumstances and AT&T's interest in not trampling its only nationwide competitor into the dust have been discussed earlier.

Finally, there was the antipathy with which the principals viewed each other. AT&T had hardly welcomed its new competitors into the communications world with open arms, nor could it have been expected to. Similarly, having just burst on the scene as heroes of the new competition but with little cash flow, the OCCs were suspicious of the entrenched monopolist. However, they desperately needed its system. In fact, they could not operate a day without it.

Nevertheless, they were convinced that AT&T would take advantage of its position, experience, financial base, and essentiality to all concerned to try to impose onerous terms of interconnection on them.

The OCCs' greatest hope was the FCC, and they had shown consummate skill in using its processes to get into the business. But the FCC was not disposed in 1978 to go through the trauma of an all-out, on-the-record access rate case of great complexity and length. The Commission had had some success in turning the parties loose to work out interim compromise solutions to their problems, and it was ready to try that again.

This was the posture in mid-1978 when AT&T gave the problem of compensating the local companies for keeping its competitors in business its best shot. It filed tariffs covering connections to competing long distance telephone services, a series of heavy documents which became known collectively as ENFIA. ENFIA stood for "exchange network facilities for interstate access," which simply meant the connection of the local telephones to the systems of the long distance telephone competitors.

The immediate reaction of the OCCs to the ENFIA tariffs—and it was expected almost totally regardless of any specific provisions—was shock and outrage. Whatever AT&T wanted, in their view, was too much (and it should be noted that AT&T's countering attitude was that the OCCs wanted complete parity at little cost in a business AT&T had spent 100 years building.)

By now, the government had provided two partially parallel and sometimes complementary communications policy-making agencies. The Commerce Department's National Telecommunications and Information Administration (NTIA) did not have the FCC's decisional responsibilities, but it was in the business of looking at communications questions that might involve the Executive Branch of the government, on rare occasions the President. Its officials were are are Presidential appointees, so their philosophical views are usually in accord with those of the Presidential designate two miles across northwest Washington—the chairman of the FCC.

At this time, NTIA was headed by Assistant Secretary of Commerce Henry Geller, a former FCC staff official and counsel with an excellent reputation among his colleagues as both lawyer and pragmatist.

Henry Geller was in the right place at the right time to give the FCC its compromise solution to the ENFIA problem. In a letter, he suggested that what was needed was a "rough justice" interim answer. He said that as a first step the Commission should "bend every effort to achieve a negotiated settlement among the parties, similar to the successful negotiations that took place under docket 20099," the late-1975 agreement.

What was called for at the time, he advised, was a "rough estimate," since the complex issues made it hard to decide on a tariff "that would place MTS/WATS and similar services on a parity with respect to using local exchange facilities, and constitute the proper contribution." The FCC accepted the recommendation.

Meeting on a heavy schedule for long hours, the participants could not even get together on semantics. They argued heatedly over a proper term to describe the amount contributed by interstate toll to local and intrastate operations to help keep local rates down. Unconvinced of the existence of any such contribution—or preferring to call it tribute—the OCCs grudgingly conceded it would have to be discussed if they could find a name for it. Looking for some meaningless and not pejorative term, they finally decided on "Ralph." (At the signing ceremony after several months of intense bargaining, AT&T handed out souvenir coffee mugs to the regular attendees. They were inscribed, "Ralph is a four-letter word.")

After weeks of proposals, counter-proposals, arguments, and indignation, AT&T and the OCCs agreed on a formula which was to apply for three years or more, subject to further FCC action. It was calculated that under the separations formula as applied to exchange plant, interstate long distance telephone services provided under AT&T tariffs were currently contributing, on a nationwide basis, 5.5 cents for each minute of use those services made of local plant. The figure was expected to go to 6.1 cents in the three years, under the steadily escalating subscriber plant factor, or "spif."

Under the agreement, the OCCs would pay 35 percent of what AT&T would pay per minute of use, as long as total OCC industry revenues were below $110 million a year. At the time, they were about $60 million. After reaching $110 million, their discount rate would be 45 percent of AT&T's, until the OCCs were grossing $250 million. Then, up to $375 million for the OCCs as a group, the payment would be 55 percent of AT&T's.

Obviously, the OCCs' discounts were the major factor making it possible for them to underprice AT&T and gain a foothold in the long distance market. AT&T apparently went along because of its concern that resisting would adversely affect its posture in the government's antitrust suit, particularly if the resistance caused the emerging competitors to go belly-up.

Until ENFIA made way some years later, at the time of divestiture, for the new access charge tariff system, it was the subject of almost continuous acrimony. At almost every turn, the carriers accused each other of misreading or misapplying the terms of the agreement or deliberately violating its terms. The terms of the agreement depended on monitoring reports of OCCs' tariffs and revenues, and questions were frequently raised as to the accuracy of those reports.

Areas of disagreement increased as the OCCs found new or different ways to offer their services and use those of the telephone companies. There were constant claims by the OCCs that they were being overcharged. All of these disagreements found their way to the FCC for resolution. During the entire period, while other digestive distress sufferers were told in television advertisements how they should spell "relief," the FCC was certain that its relief would come only with the expiration of the source of incessant trouble spelled "ENFIA."

The ENFIA controversy went on as though the question of a competitive interexchange market had already been decided in favor of the rapidly developing competition. Officially, nonetheless, the FCC was taking the position that the issues were still open, and that the market structure inquiry, docket 78-72, would be played out to a conclusion.

Shortly before the commenters in the market structure docket submitted their recommendations as to what an industry model should look like—and, as it turned out, less than six months before the Commission adopted its decision—the FCC went another step along the way to virtually unlimited competition.

By a 4 to 3 majority, in mid-February 1980, the commissioners voted to get comments on a shortcut approach to providing for shared use and resale of the switched services, particularly WATS. The four FCC members in the majority said they were taking the action out of concern that putting fully into effect the industry structure, which would finally come out of the big inquiry, would take ten to fifteen years and thus delay for that long the competition they had not yet formally adopted as a policy.

The dissenters, led by Commissioner Joseph R. Fogarty, argued unsuccessfully that the action duplicated the market structure probe and prejudged its results. The move, they contended, would inhibit an effective and timely resolution of the big issues. They strongly implied that by expediting the full-scale market entry of resellers—and it was apparent that all of the competing carriers would be resellers to some extent at least—the agency was rendering a de facto market structure ruling.

The industry models proposed in the inquiry by the interested parties—mostly AT&T, the competing interexchange carriers, government agencies, and a few other large customers—did not unduly complicate life for the FCC. AT&T led the way among the big majority of commenters with the conclusion that the basic question in the inquiry already had been decided. It recommended, in a sharp turn from traditional policies which highlighted how much things had changed, a "market structure that includes an open entry policy."

AT&T viewed the telephone companies as the only possible providers of "anybody-to-anybody" service and urged new directions to give them "increased flexibility in making rate structure and level changes."

In a conclusion not seriously disputed by any of those submitting industry models at the Commission's invitation, AT&T's Jim Billingsley said that the "initital purpose" of the inquiry "has been settled as a practical matter in favor of competition and open entry. The FCC has clearly signaled a new policy of reduced regulation [the policy has been extremely slow in being applied to the Bell companies] and increased reliance on marketplace forces. Rapid advances in technology, proliferation of new highly substitutable services, entry by well-financed, technically capable firms, and the passage of time have, for all practical purposes, eliminated the sole source option."

On August 1, 1980, in its final meeting before the annual August recess, the Commission decided in favor of what was rapidly becoming the status quo. Since few proponents of a return to the former nationwide monopoly had turned up during the comment period, the FCC was not forced to make a choice and elucidate its reasoning. It simply said it had decided against a shared monopoly in interstate services because it was convinced that competition is in the public interest and would further the goals of the Communications Act.

Somehow, the FCC did not wholly agree with those who assumed that "it would be impossible to turn the clock back to a pre-1969 era by eliminating all interstate interexchange competition." Instead, finding that evidence in favor of monopoly had not been persuasive, it volunteered, "If it had been presented in this proceeding that the net effects of competition with MTS and WATS services will be deterimental to the public interest, this Commission could take appropriate steps to restrict competition in all interstate interexchange services."

How would it do this? "Although it would not be practical to eliminate the new entrants or their facilities, it would be possible to eliminate competition without eliminating the existing competitors. The new entrants and their facilities could be integrated into a shared monopoly industry structure if such a result would serve the public interest." By deciding not to take such action, the Commission was spared the task of explaining just what a shared monopoly in interexchange service would be, or how it would work.

When enough time has passed for all of us to determine how the great experiment panned out, reliance on economic theory will have undergone its most stringent test. In shaping its policy, the FCC relied almost exclusively on economic theory, without the support of any hard empirical evidence. For example, one of the FCC's conclusions—which may or may not be true but was entirely an article of faith—was this: "Although there is no guarantee that the competition that open entry is likely to produce in interstate telecommunications services will result in an optimal mix of facilities at all points at any given time, there is good reason to believe that such competition is more likely to result in an efficient allocation of resources than a monopoly industry structure." In any event, the FCC concluded that the evidence failed to substantiate the economic theory that high concentrated volume produces lower costs, or economies of scale.

Since it had already tentatively decided in favor of a nationwide average local access charge system, the Commission said, any difference in costs to undergird deaveraged rates would have to come from differences in intercity transmission expenses. It doubted whether the cost differences would amount to enough to permit a questionable marketing strategy which would make up all losses on competitive routes by raising profits on the remaining monopoly routes.

Anyway, the FCC concluded, the monopoly routes would become few and far between as the competing carriers extended their services to reach for universal service benefits. It apparently took a better crystal ball than the Commission's economic theorists had at the time to recognize that the competitors would attain universal service off their main routes by reselling Bell System services. Thus, the greatest individual beneficiaries of Bell's uniform averaged rates are the competitors, and in large areas of the nation, interexchange service may be closer to being a shared monopoly than the Commission could have foreseen.

At one point, the FCC came about as close to a simple, direct explanation of its action as a government agency ever does. "This," it observed, "is the only realistic course open here. It would be completely incongruous for the Commission to now attempt to turn back the clock and to carve out a separate MTS-WATS enclave which alone would be the preserve of 'monopoly carriers.' "

In a more oblique manner, Chairman Charles D. Ferris put his finger on what had happened in a separate statement. "Much of the credit for this outcome," he stated, "belongs to the Execunet court for forcing us to abandon a narrow view of the MTS/WATS markets." Taking a long view in another separate statement, Joe Fogarty commented that the FCC was taking the action based on "hypotheses," and expressed fears about the future fate of nationwide averaging. The reasoning of the FCC order was "plausible," he concluded, but "the long-run results are the ones which should be of continuous concern."

One of the FCC's final directives in the market structure order was that henceforth docket 78-72 would be confined primarily to access charge matters. More than seven years after the Commission order, it still is.

For the next couple of years, the FCC struggled with both theoretical and practical aspects of a system of access charges under which the local telephone companies would be compensated for the use of their local systems within LATAs—local access and transport areas which constituted the geographical boundaries of the exchange companies' operation—for interstate interexchange service. The LATA concept came quickly after the January 1982 announcement of divestiture, but immediately prior to that time the question of compensation for local and intrastate system use provoked similar issues.

From the time that interstate interexchange telephone open entry received the FCC's formal blessing, a system of access charges to replace the short-term ENFIA arrangement had fairly high priority. Its urgency received a considerable shot in the arm, however, when the Justice Department and AT&T made their early-1982 bombshell announcement of agreement on a modified antitrust judgment, under which all the local Bell operating companies were to be divested into separate and independent entities within two years. What had been desirable—the replacement of the existing interim compensation and settlement plan with something more closely geared to competitive market conditions—suddenly became essential.

Throughout the complex debate over access charges, two issues were critical: one primarily to the interexchange carriers, and one to local telephone subscribers and those political officeholders basically concerned about them.

The interexchange competitors had their most fundamental dispute over access which they characterized as inferior, inadequate, or nonpremium. Under the arrangements in effect universally for carriers other than AT&T until the primary "equal access conversion" period of two years ending in September 1986, the competitors were tied into the line side of the local dial switching offices.

Thus, the competitors had the same connections to the local switches as the lines of all other subscribers, essentially the same hook-up as a residence telephone. Among other things, this meant that the competing carriers could not transmit onward the calls from their subscribers if they were made over conventional dial wheel telephones without some form of conversion. If the subscriber did not have a TouchTone or touch-signaling telephone, he had to employ a tone generator or pad over the dial. In any event, the user had to employ a lengthy access code for identification and billing purposes. Customers of MCI or Sprint, for example, had to dial about twice as many digits to make a call as those tapped out or dialed by others who made calls the traditional way, which meant via AT&T.

Another problem was that the line side connection denied the competitors the availability of automatic number identification or answer signaling. AT&T customers could hold the line when they made a long distance call as long as they wanted; if the called telephone did not answer, AT&T's billing equipment knew that the call was not completed and therefore the customer should not be charged. The competitors had to assign an arbitrary number of seconds to each attempted call, after which it was assumed that the call had been completed, and the customer was billed for it. Consequently, if a customer held the line longer that the initial period, but the called party did not answer, the caller was still billed for the call. The customer relations problems this caused for competing carriers were obvious.

What has come to be known as premium access thus meant a trunk side arrangement, which traditionally connected Bell toll lines. Without it, the competitors kept reminding everyone who would listen, they and their customers were second-class citizens who should pay discounted rates.

The second big area dealt with recovery of that portion of subscriber line, or local loop, costs allocated by separations to the interstate juridiction. Traditionally, those non-traffic sensitive costs were recovered entirely from revenues generated by long distance calls.

From the beginning of analysis of an access system, the FCC and most communications companies were critical of this method of cost recovery as unjustly burdening heavy users of the long distance network with costs attributable to marginal or nonusers. They leaned toward a direct assignment of those non-traffic sensitive (NTS) cost to the local subscriber. As the cost-causative user, he would pick up the tab for his local loop. The reasoning went that the local customer, whether making or receiving local or long distance calls, was the one responsible for causing the expenses of his connection to the entire telephone network.

While the proponents of the end user charges based their arguments on economic theory, the procedure would deliver practical benefits to them. The FCC and the interexchange carriers preferred it because, among other reasons, the cost of interstate long distance calls would be less if less of the NTS subscriber line costs were assigned to them. Total cost of local service for the customers of local exchange carriers would be higher, but the additive would be labeled something like "federal subscriber line charge" and clearly attributed to the distant bureaucrats in Washington, not to the friendly local telephone company.

Whether for theoretical or practical reasons, when the Commission called for comments on four possible methods of basing access charges, the system now in effect won favor with a distinct majority of the commenters. When it became apparent that the supporters of direct assignment of NTS costs —private line service costs would be separated out and assigned directly to that service—were ready to make transitional adjustments to ease the impact on local subscribers, the plan picked up even more votes in the reply comment round.

A critical adjustment, gradually worked out in several successive versions of the access charge system, was the provision of transitional arrangements so subscribers would not get the full impact of the plan all at once. An early FCC version, calling for a $4 monthly end user or subscriber line charge for most customers, provoked full outcry from the system's antagonists.

Those who opposed or at least seriously questioned the system regarded it as a threat to affordable universal service. They were interested primarily in the welfare of low income, often rural, telephone users who made few long distance calls. They included numerous members of Congress who were well aware that their constituents were more likely to be occasional long distance users than business executives who spent a good part of their days calling distant cities. Legislators, too, were frequently reminded by customer activist groups that many voters, handicapped or elderly, relied on the telephone as a lifeline to their friends and families. State commissioners, as well, knew that they had more constituents who made few or no long distance calls than they did among those who would be net beneficiaries of the plan.

The main contention of the access system's opponents was simple enough. It was, they said, favorable mainly to business, and disadvantageous to Mr. and Ms. J. Q. Public. Among businesses being favored would be the interexchange carriers, whose volumes would improve with the lower rates they could offer when not passing their local loop access costs along to long distance customers.

Complicating the debate was the obvious fact that in rural or sparsely populated areas, local loops on the average were considerably longer than in cities and consequently more expensive. If the customers of a small rural company or cooperative had to pay their own full NTS costs, their monthly rates might in truth become astronomical.

Soon, all good access plans had to have provisions for funds or subsidies to ease the burdens of high-cost areas, measured against the national average. Another requirement for any comprehensive access program was some provision to ensure that low-income users had the opportunity for affordable lifeline service.

To a large degree, the crucial political issues related to access charges took center stage in the congressional debates on new telecommunications legislation. It was the existence of the full-scale debate in Congress, and the frequent near-misses of adopted legislation, which made it obvious that only a compromise method could work. As will be the subject of more extensive treatment in chapter 17, Congress controlled the outcome by the threat of legislation about as effectively as if it had, in fact, legislated.

Feeling the heat of congressional, lobbying, and divesititure fires, the FCC shaped a comprehensive program immediately before Christmas 1982. It was trying to calculate the value of premium access to AT&T as a total dollar amount, which it concluded at that time was $1.4 billion. In fact, when the access system's time came, the interexchange carriers paid their share of NTS costs as a carrier common line rate assessed by minute of use, with nonpremium minutes assessed at 45 percent of the premium rate, or a 55 percent discount.

Still trying to find a level of the flat monthly subscriber line charge which would be acceptable to all concerned, the Commission set the rate for residence subscribers at $2 a month, and business subscribers at a minimum of $4 monthly.

The upheaval in the telecommunications industry continued during 1983, as all interested parties looked ahead, most with apprehension, to the January 1, 1984, divestiture target day. The industry struggled with a wholly new kind of tariff, led temporarily in the pre-divestiture period by personnel of AT&T.

Later, the support group was to become Bell Communications Research, Inc., commonly owned by the seven regional holding companies and carrying out a slimmed-down set of duties. The tariff and rate setting job was to go, where centralized tariffing and pooling of revenues was either desired or mandated, to a newly established National Exchange Carrier Association (NECA). But for the task of filing the first set of access tariffs in early October 1983, initially scheduled to take effect on January 1, 1984, with divestiture, the Bell companies' support group, already organized and staffed, was designated by the FCC.

At about the same time, members of the public got their first taste of what the new regime would mean to interstate long distance rates. AT&T announced that, based on its analysis and estimates of the impact of the FCC access plan, it would undertake the largest telecommunications rate reduction in history contingent on the exchange carriers' access tariffs taking effect as scheduled.

It was not all that easy. The FCC was in turn tackling the biggest tariff analysis job in its history, and it had no prior experience with the new industry structure. It took two three-month suspensions of the filed tariffs until the Commission was ready with an order permitting revised versions to go into effect early in June 1984. In the meantime, after several cliffhangers when it appeared that, at least technically, there would be no method in existence at all to compensate the carriers for their jointly rendered services, the old procedures continued pretty much by default.

The Commission slashed the cost estimates of the exchange carriers, now led by their newly formed association, NECA, left and right. It found expense forecasts too high, and demand estimates too low. The result was a cutback in message toll and WATS rates of $1.7 billion, or 6.1 percent across the board. The new regime was delivering on its promises for long distance customers.

For local customers, the outlook was different. State regulators, consumer groups, and friendly congressmen vied with each other in making the highest possible forecasts of increased local rates. To then add on a $2 monthly subscriber line charge, they argued, would be the last straw for many poor and elderly, who would have to give up their lifeline.

Small businessmen joined the protesters, contending that many of them made few long distance calls and could not afford the added flat-rate cost imposed on their business lifelines. Warned by strongly expressed sentiment in the Senate in favor of legislative prohibition of end user fees on residential and single-line business subscribers, the FCC took what was considerably more than a hint. It deferred subscriber line charges for those customers for a year. The only end user fees to be paid until June 1, 1985, would be $6 a month per line by multi-line business customers.

While a federal/state joint board took up the subject of future access charges, the FCC received one of its most ringing endorsements from a federal court. Numerous and wide-ranging appeals to the access charge orders had gone to the U.S. Court of Appeals for the District of Columbia. The points of attack, and therefore the issues to be argued, were so numerous that the court, in an unusual step, set aside two days for oral argument instead of the customary one or two hours.

In mid-June 1984, the tribunal gave its answer in a 110-page printed opinion, of which the first seventeen pages had to be devoted just to listing the cases at bar and the attorneys representing the parties. The FCC's access charge system was almost completely upheld and remanded to the Commission only in two areas of relatively limited application. The court directed another look at access charges for telephone party lines and the treatment of small companies receiving their toll settlements on an average schedule basis.

The judges' approach was one taken when a court finds nothing serious to object to in an agency decision on appeal. In a sentence that might have astounded observers of the Execunet controversy, the tribunal announced, "We are not a policy-making body." (Cynics might reply, "Except when it chooses to be one.") The access charge court continued, "This court instead patrols the perimeters of an agency's discretion. If any agency in the course of an informal rulemaking does not attempt to close itself off from informed opinion or to extend its reach beyond the scope of permissible authority, then it is our duty to accept that judgment if it is rational and not unreasonable."

Once again, the Supreme Court declined without comment to review the appellate tribunal's decision, and the FCC's access charge procedure had passed one of its two big tests. Its proponents had only to tread fairly softly to bypass the dangers of a sometimes unruly Congress.

The final compromise in this series came in mid-November 1984 from the federal/state joint board and was, hardly surprisingly, ratified by the FCC a little more than a month later. It called for a $1 monthly end user charge, half of the previously projected subscriber line fee, to be effective June 1, 1985. After a year, it was decided, the charge would move up to $2 and then be capped pending a further review of the access system.

There was clear indication that the plan had been checked out with leaders of congressional committees before it was formally adopted, and the legislators immediately issued statements accepting the transition plan without any serious objection. Only consumer organizations took the opportunity to publicly voice their dissent in a demonstration which was described by a leader as "the first of its kind."

Pickets from organizations including Public Citizen, Congress Watch, the Gray Panthers, and the Consumer Federation of America swarmed the hallways outside the Commission's main meeting room as the vote was taken. They carried placards, and some used megaphones to shout messages to FCC Chairman Mark S. Fowler. The principal message was one which demonstrated again that the now-dismantled Bell System still carried for some the label of the century-old dominant monopoly.

"Fowler, Fowler, you can't fool me," demonstrators shouted. "You're in bed with AT&T."

16

ALWAYS THE DEFENDANT

For a century, the regulatory regime under which federal and state authorities controlled market entry and purported to oversee the rates and practices of monopoly telephone companies was the companies' principal shield against antitrust actions.

In the years immediately before World War I, AT&T had warded off threats of antitrust litigation which came from rough-and-tumble competition at the local exchange level by letting the competitors in effect have their own piece of the status quo. Instead of buying out or forcing out local telephone companies operating in the same places that the Bell companies had systems, AT&T agreed in landmark concession to keep away from others' local turf unless called to serve there by—guess who?—the regulatory authorities. The concessions were named for two company executives, the Hall memorandum and the Kingsbury commitment.

Nor would AT&T any longer pressure non-Bell telephone companies to sell out by denying them access to AT&T's long distance lines. Otherwise, Bell and independent telephone companies had their own franchised local enclaves, separate and more or less equal. Both enjoyed or at least tolerated the controlling, and protecting, hand of the regulators.

When the federal government filed the 1949 antitrust complaint, for reasons centered on the Bell System's virtual monopoly in manufacturing and procurement and not targeted on the local telephone monopolies, except as purchasers, again it was the system of regulation that came to the rescue. As recounted in chapter 6, the FCC convinced the Justice Department that the regulators had effective control of Western Electric Co. prices which wound up in the telephone rate bases. And the tariff solution was to confine the Bell companies exclusively to those services offered under tariffs filed with the regulators.

During most of the Bell System's first century, the system had worked as envisaged by AT&T's principal architect, Theodore N. Vail. Vail, a genius at least by nineteenth- and early twentieth-century standards, was AT&T president for two separate and crucial terms. He gave the Bell System the rules of the road by which it lived and prospered until it was confronted with external economic and technological developments of a changing tele-communications world which were beyond its ability to control.

Vail's persuasiveness and prescience were incredible. He satisfied govern-ment authorities, customers, and future generations of Bell System managers that the public, even more than AT&T stockholders, was best served by monopoly telephone companies operating under government surveillance and semicontrol—dedicated to the proposition that if the com-panies met their service obligations, they and their owners would prosper as a direct result. (Modern cynics who might be tempted to call this a version of the trickle-down theory are reminded that the public *did* get good service at comparatively low rates, and that AT&T management and stockholders received stable, adequate, but hardly inflationary rewards, and definitely did *not* cut any melons.)

Vail had another guiding principle which served the Bell System and the public well until it collided with modern economic realities and antitrust policies. It was that the telephone companies provided an end-to-end service using whatever equipment was necessary; they did not lease telephones or other equipment. Subscribers were not renting a group of disaggregated pieces of hardware; they were paying for a service enabling them on demand to talk to whomever they wished to talk.

All of this worked reasonably well (in the view of many, very well) until, slowly and increasingly, competitors to the Bell System appeared on the scene. As the natural monopoly concept of telephone service became sub-ject to increasing and successful challenges, the fledgling competitors were encouraged not only to protest to the regulators, but to file antitrust suits. Tom Carter's example was followed, both at the commissions and in the courts, even if the antitrust judgment he won was small by later standards.

The more than forty private antitrust suits, which at one time were pen-ding against the former Bell System companies, were based on claims that could be divided into three categories. Most numerous were complaints by suppliers of various devices which could be used in conjunction with tele-phone service or substituted for telephone-company-supplied terminal equipment. They contended that AT&T's unlawful refusal to accommodate to this emerging new world of supply and demand, its opposition to equip-ment registration or certification, and its insistence on protective connecting arrangements either pushed them out of business or severely curtailed their sales.

Second in number were the allegations by interexchange competitors that AT&T had followed a variety of anticompetitive practices, especially artificially low pricing of its own services and interconnection restrictions imposed by its "bottleneck" local exchange companies, to try to force them out or sharply limit their market penetrations.

Finally, there were a few contentions of monopolized purchasing, or monopsony. For example, ITT won a substantial settlement out of court by contending that AT&T's role as almost the only significant purchaser of telephone equipment in the United States, and its preference for gear made by its subsidiary Western Electric Co., had imposed major limitations on ITT's market for advanced switching equipment.

As a completely integrated supplier of both telecommunications equipment and service, the Bell System touched everyone in any part of the communications business. Its sheer size made it a frightening competitor, one which had it in its power, unless reversed by governmental authority, to lock the door even on market entry itself. The Bell System was so pervasive that almost everyone who failed in a communications-related business, or did not do as well as was originally expected, was convinced that AT&T was at the bottom of the problems.

The Bell System's huge size made it something else too—an attractive target for such litigation. Big money was there for any complainant who was even partially successful, making the cost of undertaking and prosecuting an antitrust suit an attractive gamble. Further, added credence and justification were given to these claims of wrongdoing by Bell when the federal government joined the private litigators in 1975 with its own antitrust action. Moreover, the vast discovery projects of the Justice lawsuit and the potential findings in the government's case limited the private complainants' legal costs and improved their chances for success.

Always the defendant, AT&T could not win one of the suits. Even when AT&T was successful in gaining complete dismissal of the pending charges, as it was on a few occasions, the antitrust cases each cost millions to defend. It was therefore tempting to discuss multi-million-dollar settlements of the pending cases, despite some large payouts, to cap the legal expenses and avoid the risk of much larger judgments.

There have been no published estimates of the cumulative cost to the former Bell System partners of the private antitrust cases, and there can be none, for two reasons. First, the parties to the majority of out-of-court settlements agreed not to disclose their terms. Second, while the burden of direct legal expenses could be estimated, there is really no way to calculate the impact of the trials on the whole organization, not only in dollar expense, but in time spent and lost, and disruption to normal business. Some published estimates of the cost to AT&T of defending against the government suit may seem huge, but to informed observers, they appear very conservative.

Just adding the out-of-pocket cost of the known or estimated judgments and settlements in the private cases runs to something approaching a billion dollars, even without the interest on judgments during the appeal period, another big cost. The litigation must have cost the Bell companies altogether, in direct dollar outlays which were not productive of any revenue-producing service, far in excess of another billion dollars. The total burden of antitrust litigation, public and private, in dollars, time, effort, and disruption, was so enormous that divestiture as a means of bringing it to a close had to have been a tempting prospect. It should be remembered that generally regulators view the cost of losing antitrust defenses, especially the costs of settlements and judgments, as below-the-line expenses not subject to recoupment from a carrier's rates.

In March 1980, when the flood of private antitrust suits was at its peak, AT&T reached a settlement in the case brought by the Wyly Corp., parent of the then-defunct Data Transmission Co., and Datran's trustee in bankruptcy. Datran's troubles were recounted earlier, in chapter 12. Settlement was for $50 million, with Wyly expecting to get about $25 million after payment of attorneys' fees and litigation costs, and payment of about $11 million to the trustee to settle all outstanding claims against Datran.

AT&T Vice President William G. Sharwell, in a statement issued at the time of settlement, remarked, "One of the facts of life is that our size makes an attractive target for litigation" but warned that "you can be sure we don't intend to be patsies." But, he added realistically, "Preparing our defenses in several major antitrust cases simultaneously is obviously a considerable burden and expense. It only makes sense to do what we can to clear the decks."

When contemporary events directly relevant to it are considered, perhaps the most tortuous course of any of the private antitrust suits was taken by the complaint of Litton Systems that the Bell companies were guilty of monopolization in restraint of trade by engaging in predatory and anticompetitive conduct and conspiracy to maintain an unlawful monopoly. Much was made of the PCA (protective connecting arrangement) issue and AT&T's position as stated to the FCC during the post-Carterfone period against equipment registration or certification.

In terms of its final outcome, the Litton case was the most costly to the Bell companies of all the private suits. It also was, in the opinion of some close observers and participants including Strassburg, a miscarriage of justice. Strassburg has made his position clear in testimony as an AT&T witness in a case argued along similar lines by Phonetele, Inc., after having been called by MCI in its antitrust case and being treated by AT&T as a hostile witness in the Justice Department case.

He said of the post-Carterfone tariffs, "AT&T was making a good-faith effort at the time to accommodate the network, a national resource, to these new uses, and that they could no longer be expected to be all things to all people as they had been for generations in the days of plain old telephone service." Further, he went back to the recording devices case, as recounted in chapter 3, to note that the FCC prescribed or ordered essentially the same kind of PCA later under attack.

The Litton case started off with good omens for AT&T's consistent argument that pervasive regulation insulated it from antitrust punishment, because all of its actions were taken under government supervision. It had filed a motion for a directed judgment against Litton, and U.S. District Judge William C. Conner referred the motion to Magistrate Ken Sinclair, Jr., for a report. Magistrate Sinclair came back with the recommendation, in late September 1979, "This action should be dismissed on grounds of implied immunity from the antitrust laws."

Sinclair told Judge Conner, "The continued pursuit of this action constitutes a contemplated retroactive application of the antitrust laws to conduct previously gauged by regulatory standards which are predicated on concepts of public interest inconsistent with the pure competitive model enforced by the Sherman Act. This conflict, and the comprehensive regulation embodied by federal and state programs, require that the action be terminated."

Alternatively, he suggested that if his principal conclusion was rejected, the matter be referred to the FCC under the doctrine of primary jurisdiction. That doctrine was in vogue in the early days of antitrust suits such as Tom Carter's, which in fact made its landmark contribution as a result of such a referral. More recently, federal courts generally have refused to concede that the FCC, where antitrust violations are charged, really has primary jurisdiction.

Within little more than a month, at the beginning of November 1979, a directly opposite conclusion was reached by an authority carrying more weight, the U.S. Court of Appeals for the Third Circuit in Philadelphia. Dealing with another complaint against AT&T, which was later to be settled, by Essential Communications, Inc., of New Jersey, the appellate court ruled that neither the Communications Act nor the state regulatory laws "provide a basis for an implied exemption" from the antitrust laws. It overturned and remanded a lower federal court dismissal of the case on the basis that "the conduct alleged in the complaint was exempt from antitrust scrutiny because it occurred in an industry and an area subject to regulation by the FCC."

By the following March, Judge Conner was ready to rule on the pervasive regulation argument in the Litton case. He reversed the magistrate's finding and ruled that AT&T did not have antitrust immunity. On the magistrate's alternative suggestion, the court concluded that the case was not a proper one for referral to the FCC under the doctrine of primary jurisdiction. Judge Conner commented, "The FCC has already employed its expertise in deciding the principal issue with respect to that for which that expertise was needed: whether the interface device requirement was fair and reasonable."

Since the Litton case was being tried in New York, it was inevitable that the ultimate conclusion would go to the Court of Appeals for the Second Circuit, also located in Manhattan. Slightly more than a month before the jury in the Litton case came in with its verdict, a three-judge panel of the appellate court issued a decision favorable to AT&T on a similar set of facts and allegations.

It reversed a jury award of more than $16.5 million to a Connecticut interconnect company—a firm leasing and selling terminal equipment, usually to businesses—the Northeastern Telephone Co. The judges unanimously held the following:

> Northeastern has not directed our attention to any unequivocal evidence of [Bell's] specific intent to monopolize or unreasonably to restrain trade. Instead, it points to documents circulated among the corporations of the Bell System and to testimony of Bell employees.
>
> These show only that appellants [AT&T] wanted to win the competitive struggle. This desire, without more, is not unlawful. The crucial question is whether appellants specifically intended to vanquish their opposition by unfair or unreasonable means. Northeastern's direct evidence is insufficient, by itself, to enable the jury rationally to conclude that appellants possessed the requisite intent.

Litton's damage claim was far larger than Northeastern's, and at this point AT&T's position in the line of cases based on the post-Carterfone tariffs' PCA requirement and the company's objections to equipment registration or certification was looking fairly strong. From that time on, however, much of the news was bad for AT&T.

For starters, the Litton jury concluded, in a decision reached immediately before 1981's July Fourth holiday weekend, that Litton should receive $92 million of the $570 million in damages it was contending for. Tripled, in the manner of antitrust judgments, the Bell companies would be liable for $276.6 million plus interest during the intervening period, if the jury verdict was upheld on appeal. It was found that the Bell companies willfully maintained monopoly power by predatory or anticompetitive conduct, but were not guilty of conspiracy to monopolize.

About a month later, AT&T President William M. Ellinghaus was called as a witness in the Washington trial of the Justice Department's antitrust complaint, going on at the same time. Bill Ellinghaus may have been the most personally popular telephone executive in history, among people who not only liked but trusted him. Certainly, we have never heard anyone ever question his intergrity. He was highly visible, appearing frequently as a witness in Washington and having been sent to the New York Telephone Company as president to pull that company out of the severe service crisis of the early 1970s.

He testified to what we feel, based on personal experience and close observation, has to be believed—that the decision to require PCAs was based on a desire to protect the network from harm and was never intended as an economic barrier by adding another cost to users of non-telephone company terminal equipment. He said the AT&T officials relied on the opinion of the company's technical experts that there "was a serious risk of harm to the safety and quality of telephone service if the indiscriminate interconnection of customer-provided terminal equipment were permitted."

During the Litton trial, there had been references to allegations, which in a few cases resulted in admissions or convictions, that certain Litton Systems officials had made gifts to procurement officials, allegedly including the services of prostitutes, in exchange for their patronage. After the jury trial, there remained one item of business for Judge Conner in dealing with the plaintiff which was ultimately to receive the largest judgment or settlement of all the private antitrust litigants.

The jurist ruled that Litton was not entitled to trial costs and attorneys' fees running into the eight-figure range—in other words, over $10 million. In doing so, he found "gross negligence and intentional misrepresentation" by Litton's attorneys, charged with intentionally withholding documents from AT&T's representatives during the discovery period. The attorneys presumably received their fees, but Litton paid them, not AT&T, as it would have under the usual practice.

Another panel drawn from the same Second Circuit which earlier had reversed the jury award to Northeastern Telephone found, on appeal, that everything was fine in the Litton trial. The only solace to AT&T was that the appellate court upheld Judge Conner's ruling that Litton had to pay its own attorneys' fees and costs.

Much of the appellate tribunal's opinion was devoted to the Noerr-Pennington doctrine, which holds that an antitrust defendant cannot be held liable for acts of legitimate advocacy of public policy. AT&T argued consistently—and such jurists as Judge Harold H. Greene in the Justice Department case agreed with it—that its support for protective connecting arrangements, or PCAs, and its opposition to equipment registration were exactly that. The Second Circuit disagreed.

But the crucial factor which worked against AT&T in both the trial and appeal of the Litton case was the FCC's ruling in 1976, when prescribing its registration certification procedures, that the PCA tariff was unnecessarily restrictive, unreasonably discriminatory, and therefore unlawful. That ruling would appear to be somewhat unseemly, considering that the FCC had accepted the PCA tariff as consistent with its Carterfone ruling and its policy to protest the network.

In his Phonetele testimony, Strassburg pointed out to the court the context in which that apparent cheap shot by the FCC at the PCA tariff had to be considered. It had to be kept in mind, he said, that it was only when the FCC was finally in the position to prescribe and administer a registration program of its own in order to protect the network that it no longer regarded the PCA as necessary. (Obviously, AT&T, although it opposed a registration program for a myriad of reasons, including some of dubious merit, was in no position to unilaterally initiate and administer any such program as a disinterested party.) Moreover, even with a governmentally administered registration program now in place, the FCC chose not to order AT&T to cancel the PCA requirement, but permitted it to remain fully effective as an alternative to the customer's use of registered equipment.

Things were deteriorating rapidly in this area for AT&T. The day after AT&T asked—unsuccessfully once more, as it turned out—that the Supreme Court review the Litton case, a U.S. district judge in Washington, Thomas Penfield Jackson, ruled that based on the Litton verdict, "collateral estoppel" could be invoked on behalf of a group of complainants in a class action suit headed by Jack Faucett Associates. Collateral estoppel meant that the issues of fact presented by the class action had already been tried and resolved in the Litton case. All that remained to be litigated was the amount of the damages that complainants suffered by reason of AT&T's already proven misconduct. Other antitrust complainants were surfacing around the country also seeking to piggyback on Litton, and the Faucett case was a class action suit open to anyone who had used a PCA between late 1970 and mid-1978.

Soon thereafter, AT&T counterattacked. It asked the U.S. Court of Appeals in Washington to reverse Jackson in the Faucett case. Events had proved, it declared, that the FCC's "pro-competitive policies are now undermining the longstanding policy of universal telephone service in precisely the way John deButts and many state regulators feared they would."

Events since the end of the Litton trial, AT&T went on,

have proved that AT&T and the state regulatory commissions were correct in asserting that competition in telecommunications would cause local telephone rates to rise to the detriment of the general public.

There can no longer be any substantial doubt that these facts are true. Although there has been a great deal of controversy about the economic effects of competition in telecommunications over the past 15 years, the existence of a substantial adverse impact upon the ordinary user is now an established fact. The historic rate structure in the telecommunications industry, under which local rates were subsidized by contributions from other services, is a thing of the past: local rates are skyrocketing, and the FCC has admitted that it can no longer prevent such rate increases through changes in the separations formula. Judge Greene has expressed profound concern, blaming the FCC. Congress is considering several pending bills, and others are soon to be filed."

The next year, 1984, was not only the first year of divestiture; it was a "win some, lose some" year for AT&T in private litigation. In January, to a sense of déjà vu by industry observers, the Supreme Court declined to review the Litton case without comment. The defendants—now split up and represented in these matters by a committee, while each regional holding company paid about 10 percent of the tab, and AT&T paid 30 percent—had to pony up the treble damage award of $276.6 million, now swollen to $340.8 million by interest accumulated during the review period.

But in an area of potentially higher risk, the ruling that collateral estoppel could be invoked in the Faucett case was struck down. The Court of Appeals in the District of Columbia accepted only one of Jackson's findings, that class-action plaintiffs had the right to sue. In the process, the court was highly critical of its Second Circuit colleagues' rulings in the Litton case, especially a finding that Judge Conner's refusal to accept evidence on state commissions' acceptance of PCAs was an error, but a harmless one. The federal judges in Washington found considerable harm, but their sharp words had no effect on the outcome of the Litton case.

In the series of private antitrust suits against the Bell System in the decades of the 1970s and 1980s, two others bear mention. One demonstrates, just as other cases recited in this chapter do, the importance of the background and beliefs of the judge who is assigned to the case.

The Southern Pacific Railroad Company contended, as had MCI Communications Corp. earlier, that AT&T's predatory pricing and other anticompetitive practices kept it from getting a good start in the specialized private line service field. Operating as Southern Pacific Communications, or "Sprint," it came in later—as did MCI and the ITT subsidiary, U.S. Transmission Systems—with second-round suits alleging that AT&T also had violated the antitrust laws when its competitors began their forays into plain old long distance telephone service.

Judge Charles R. Richey, a former member of the Maryland state commission staff, was a true believer in the pervasiveness of regulation. He felt its public interest and statutory values were different from and took precedence over those of the antitrust laws. He made his opinions clear during the trial. When it was over, near the end of December 1982, he dismissed the Southern Pacific suit, reaching all conclusions in favor of the Bell System defendants.

Southern Pacific Communications had been asking for $229.1 million after having cut it down twice from an original asking price of $567 million but Judge Richey's answer was "zero." He went back to some of the earlier court conclusions on the subject to find that "all the rates and practices of [the Bell companies] challenged by SPCC in this case are subject to pervasive federal and state regulatory control under a public interest standard that is quite different from and inconsistent with the application of the antitrust laws."

Later, Judge Richey's decision was upheld on review by higher courts. The Court of Appeals panel deciding the case took a fairly dim view of some of Richey's comments and rulings during the trial and his liberal, verbatim use of AT&T's proposed findings, but found no legal basis to overturn his verdict.

Greatly different in outcome was MCI's first suit against AT&T, covering allegations of improper and illegal AT&T conduct from the time the complainant entered the specialized private line field until its full-scale entry into the Execunet business. MCI's attorneys had leveled twenty-two counts of unlawful activity against AT&T.

The complaint was filed in March 1974, shortly after the events involving foreign exchange and common control switching arrangements (FX and CCSA) and the Bell companies' pulling the plug on MCI customers, as recounted in chapter 13. The trial—held before a jury requested by AT&T, reportedly on the basis of John deButts's belief that the AT&T story should be told to members of the public—got under way just over six years after the complaint was filed.

AT&T moved for a directed verdict, and U.S. District Judge John F. Grady narrowed the number of counts on which the company had to stand trial to fifteen from the original twenty-two. In the process, he commented that he was giving MCI "the benefit of every doubt."

Judge Grady dismissed charges that the Bell companies conspired with each other; that the Bell System increased its capacity to handle data communications to destroy potential competition; that series 11,000 (a short-lived bulk rate private line offering) was introduced to discourage potential customers from dealing with MCI; that AT&T disparaged MCI; that AT&T conducted a massive propaganda campaign against MCI; that the Bell companies brought "sham proceedings" before administrative and judicial bodies, the plaintiffs' counter to defendants' invocation of the previously discussed Noerr-Pennington doctrine; that joint telpak—by AT&T and Western Union—was intended to be anticompetitive; and that AT&T induced Western Union to file its tariff offering the same reduced rates as MCI between Chicago and St. Louis.

Slightly more than two months after the trial began, the jury came back with the verdict which shocked the American business community and certainly its telecommunications segment. Finding AT&T guilty of ten of the fifteen counts being tried, the jurors apparently applied the same arithmetic to MCI's damage claims, and allowed two-thirds of the $900 million being sought. Trebled, that meant an award of $1.8 billion, the largest in private antitrust history until the $10 billion initially handed down in the Texaco-Pennzoil case.

One crucial ruling by the typical blue-collar, Chicago urban jurors was on an issue that specialists had spent years studying and debating without being able to come to a consensus. Judge Grady declined to give any guiding instruction on the issue, leaving it entirely to the jury's wisdom. Average, fully distributed, or embedded cost was the appropriate standard to be used in measuring whether AT&T's pricing was predatory, rather than marginal or incremental cost, so the jury confidently assured all of the earnest regulatory economists who had spent many hours arguing inconclusively about a question which resembles some of the all-time classic dilemmas.

The difference between the regulators and the jurors, of course, was that the latter had the power to impose their answer to an abstruse economic question, apparently reached on a "David and Goliath" emotional level, on AT&T at a potential cost to the company's shareowners exceeding anything ever before contemplated.

The listing of the counts on which the jury decided nearly spells out what the case was all about. The jury found AT&T guilty of anticompetitive actions on the following counts: refusing FX and CCSA interconnections to MCI; tying local service to AT&T long distance service; interfering with MCI customers by disconnecting FX, CCSA, and other services; denying interconnection for multi-point service; and denying interconnection beyond a defined distance from MCI's terminals.

Also, it concluded that AT&T was guilty on charges of providing inappropriate or inefficient equipment or procedures for interconnection; negotiating in bad faith for an interconnection agreement; filing state tariffs in bad faith; predatory pricing of hi-lo, its private line voice tariff which got away from nationwide rate averaging by setting up "hi" and "lo" rate categories; and preannouncement (before the actual tariff filing) of hi-lo.

Not guilty conclusions came on five other counts: discriminating against MCI and in favor of Western Union on interconnection; charging MCI unreasonably high prices for interconnection; late or faulty installations; interfering with MCI's financing; and predatory pricing of telpak.

Judge Grady commented in a colloquy with AT&T counsel after the verdict came in, "I think that the weakest part of the plaintiff's case is the damage study," described as a "lost profits" study. But, the judge went on, "It was either direct a verdict or let it [the study] go." In any event, AT&T lost no time taking the case to the Court of Appeals for the Seventh Circuit in Chicago.

With both sides on tenterhooks waiting for some disposition of the huge damage award, on which interest was mounting rapidly, it was more than ten months, in early May 1981, before a three-judge panel of the appellate court heard oral argument. "Sweating it out" continued for another ten months, until late March 1982, when the court announced that one of the three judges who had heard the argument was ill, and that the case would have to be reargued.

It was mid-January 1983, nearly three years after the jury trial started and close to nine years after the MCI complaint was originally filed, when the reconstituted panel came out with its verdict. The decision was 2 to 1 on crucial questions of predatory pricing and rejection of the lost profits study, making it obvious why the original panel needed its third member to reach a decision.

On the scorecard of counts, the appellate tribunal sustained the jury's guilty findings on seven of the original twenty-two. It overturned the two jury findings that AT&T's handling of hi-lo was predatory, as well as the count related to denial of multi-point interconnection. Left standing were the five counts on interconnection and the two dealing with filing the interconnection tariffs with the states. It held the trial judge to have erred in not directing the jury to accept incremental rather than fully distributed costs as an acceptable measure for competitive pricing.

As awesome as the $1.8 billion verdict by the jury had been, by the time of the appellate decision, it had become about $2.1 billion, with $300 million in interest added. Before the end of 1983, after the usual Supreme Court refusal to review despite requests by both sides, the case was on its way back to the District Court in Chicago. MCI had won guilty verdicts on seven of its twenty-two charges, but the measure of damages, fully distributed costs and the lost profits study, had been struck down.

In this greatest legal confrontation of telecommunications history (others may put the Justice Department suit in that position, but it was not carried through to the finish, and eight viable entities were left where there had been one before), several fundamental things had changed dramatically in the past several years. Meanwhile, AT&T also brought in a new team of trial lawyers.

Testimony in the new jury trial, this time strictly on the amount of damages, did not begin until April 1986, only two months short of five years after the first jury had handed down its cataclysmic award. The first jury had been given a perception in which the Bell System was a colossus, with deep pockets able to afford a gargantuan award on behalf of the struggling little MCI, striving to stay alive in a dog-eat-dog business world. In the first case, AT&T was a pit bull and MCI was a friendly little toy poodle.

Jurors do not live in vacuums, and these jury members were well aware that AT&T was well over two years into divestiture. Now, AT&T had less than 30 percent of the former Bell System's assets, and its earnings reports recounted on the financial pages were suffering considerably. Judging from the composition of the new jury, much different from the first one in occupations and backgrounds, some of its members probably did read the *Wall Street Journal* or at least the business section of the *Chicago Tribune*.

There was another, perhaps subtler, change in the atmosphere. Although the case supposedly covered the period before message toll telephone competition, this was a time of equal access programs and calls on subscribers to select their preferred long distance carrier. On the ballots, MCI and AT&T were ranked equally. The members of the jury presumably all had seen extensive television advertising in which MCI and AT&T went head to head, slugging it out as equals.

MCI and its counsel, who successfully presented the image of the poor struggling underdog in the first trial, may have made two other, if related, tactical errors. They let their economic witness, William Melody, perhaps the most omnipresent courtroom opponent of AT&T in history, come up with a new damage study which led to a claim that dwarfed the earlier $1.8 billion award. This time, MCI was asking $5.8 billion, before trebling, instead of the first trial's original $900 million demand.

Second, Bill Melody tied Execunet into his damage study. He made the somewhat speculative claim that Execunet could have been started much earlier if not for AT&T's anticompetitive actions. In any event, even though MCI lawyers argued that there was no conflict with their position in the first trial, that damages were not being sought for Execunet, but that the question would be litigated in a second suit (MCI II), it evidently left jurors wondering whether MCI had not in fact changed the thrust of its case and was now overreaching.

After hearing five weeks of testimony, the jurors concluded at the end of May that MCI was entitled to $37,775,920, or $113,327,760 after trebling. No small change, but only a little more than 6 percent of the original award. More than five years of interest sought by MCI, from the date of the original award, would have raised the tab considerably, but Judge Grady ruled that interest should run only from the date of the new verdict.

After the extended litigation, the case now appeared ripe for settlement, especially in view of another ruling by Judge Grady. Each of the Bell regional companies, having received roughly 10 percent of the assets of the former consolidated Bell System, was liable for about 10 percent of the total award. But MCI had reached settlements with two of the regionals, one before the jury trial and one immediately thereafter.

Terms of neither were announced, but U S West settled both pending cases for "business arrangements" and cash which, based on a later MCI financial report, was at least $63 million. Bell Atlantic, in a different type of settlement, obtained a "floating line of credit" against its antitrust liabilities with MCI for what was reported later to be $30 million.

AT&T contended that the other defendants should receive credit for the value of the two settlements against the total award, and Judge Grady went along with that contention. MCI strongly objected, and a week later the jurist certified the question to the Court of Appeals. The two parties also were trying to get together on the very considerable matter of attorneys' fees and costs, with AT&T arguing that it should not pay fees and costs on the counts for which it was found not guilty.

Thus, with MCI II also going through a complex and expensive discovery process, both parties were still very much at risk. It came as no surprise in late November 1986 that the two announced they had settled both suits. The amount was not announced, but evidently was somewhere between the jury award of $113,327,760 and a maximum of about $165 million. The maximum was extrapolated by a security analyst who was told—correctly, as it turned out—that AT&T would not separately report the amount of its payment as not material under accounting rules. For that, AT&T's 30 percent share of the tab had to be less than 5 percent of the preceding year's net income, or under $50 million.

It is unusual for a litigant to be happy when settling a case for a nine-figure amount. This time, however, under all the circumstances, AT&T's lawyers made it plain that they had celebrated what they considered a victory. MCI, which had come close to a bonanza in the billions, was clearly a bit disappointed.

17

THE BELL BILL AND OTHERS

To the management of AT&T, determined to preserve a business way of life that it was convinced was best for the nation and its people, Congress was something of a last resort. The telephone people felt more comfortable in the familiar setting of a regulatory commission, but the FCC was steadily assuming a more pro-competitive stance. There was no reason to believe, certainly not with any assurance, that the courts would overturn what was going on. In fact, the evidence was to the contrary.

The telephone industry had never before played the lead in a big legislative drama. Some telephone people were no strangers on Capitol Hill, but usually their appearances were in connection with general legislation, such as tax bills, or in prominent supporting roles, for example, with the communications satellite bill. They had testified in defense of what they had done, such as in the hearings shortly following the 1956 antitrust consent decree, or to oppose what seemed to be fairly remote threats of restructuring, as exemplified in a proposal in the 1960s by a prominent liberal, Sen. Philip Hart (D., Mich.).

Senator Hart, who gained such respect from his colleagues that they named the newest of the Senate office buildings after him, headed the Senate Judiciary Antitrust and Monopoly Subcommittee. By the time he reached telephone industry witnesses, he had been conducting hearings, off and on, for a year on a bill to set up an industrial reorganization commission and court. The purpose of the proposed new organizations would be to restructure, and make more competitive, seven basic U.S. industries. They included the telecommunications industry.

Defending what they considered the very heart of their business before Congress for the first time, AT&T officials presented a mass of material to the subcommittee. It included a cost study concluding that residential service rates would have to be increased 75 percent to meet the direct costs of providing service, if their industry were restructured to separate local and long distance operations.

The appearance before the Hart subcommittee was at the beginning of August 1974, at the time when competition was just getting a small but firm foothold in the communications industry. Senator Hart's proposal never reached the stage of serious consideration, and probably would not have, even if his health had not been seriously declining. But the hearings did serve to remind AT&T management that, with things not looking so good in the commissions and courts, there was another, and very influential, branch of the government to address.

AT&T management remained convinced of the rightness of its cause, one articulated most strongly, as previously observed, by Chairman John deButts. In 1975, the decision was made to draft legislation, seek support from groups with similar interests, and march in large numbers from the grass roots to Capitol Hill.

The grass roots strategy gave the proponents of maintaining the status quo through legislation fairly considerable lobbying muscle. The Bell companies and mostly like-minded independent telephone companies operated in every community in the nation. Telephone unions were equally concerned about the potential erosion of their employers' business, and their members were all over the country. Many state regulators had a similar view of the situation. All told, there was hardly a member of Congress who did not have campaign workers and contributors, college roommates, or civic club associates among the ranks of those opposing telephone competition.

To someone relatively naive about the congressional process, as the AT&T management proved to be, it evidently appeared that lining up enough committed supporters would be sufficient to carry the day. On that basis, a legislative wish list was drawn up, and plans were laid to urge a large number of legislators to introduce it in bill form.

By early December 1975, it was reported in *Telecommunications Reports* (and nowhere else, to the future puzzlement of those who saw divestiture become the big business story of the decade), the Consumer Communications Reform Act, or CCRA, was in final draft form. The bill purported to reaffirm congressional intent of the existing statutory scheme, but actually it would have made drastic changes. In effect, it would have ratified the monopoly character of the industry which had been in place before the FCC approved competition in terminal equipment and specialized intercity services. It would transfer authority over terminal equipment to the states, which were generally opposed to competition. In the case of intercity services, the bill would have precluded the FCC from licensing competitive entry in any case where existing carriers were capable of providing an existing or new service.

The work continued without further publicity and did not hit the daily newspapers until John deButts referred to the proposal in a talk to security analysts in early February 1976. At that point, with the campaign now in the kickoff stage, outlines of the proposed bill were distributed generally to the press.

The outline included the following:

In the proposed bill, Congress would state that the integrated system of common carrier telecommunications services is an essential element in achieving reasonableness of charges and universality of service. Accordingly, Congress would reaffirm its policy that the integrated telecommunications network should be structured so as to assure widely available, high quality telecommunications services to all of the nation's telecommunications users at the lowest possible cost. Conversely, Congress would find that authorizations designed to foster a multi-supplier environment for interstate services are contrary to the public interest.

The step-by-step, pound-the-hallways lobbying job began. While the telephone companies and their supporters started lining up congressmen to drop the bill into the House and Senate hoppers, the competitors began developing their response. With relatively limited resources, the new intercity carriers formed a committee, the Ad Hoc Committee for Competitive Telecommunications, or ACCT, put a public relations firm to work, and generally began their grand alliance with existing terminal equipment, computer, and other industry associations against what was very quickly dubbed the "Bell bill."

Without public comment or notice, Rep. Teno Roncalio (D., Wyo.) dropped the CCRA into the House hopper March 4 to become the first of many to introduce the bill. Some two months later, a couple of versions of the CCRA had ninety-six sponsors in the House and about a dozen in the Senate.

Interestingly enough, a legislative proposal by the FCC of a largely procedural nature was moving quickly through the Congress at this point, the last time that anything related to telecommunications legislation would engender anything except a major battle. The bill lengthened the time the FCC could suspend a tariff filing and the notice period that could be required before the filed tariff took effect. It was a reaction to AT&T's recent filing of a series of major new tariff proposals, mostly in the private line area, and all concerned needed more than the usual time to consider them. The bill became law without opposition.

By early August 1976, 157 members of the House had introduced the CCRA in bill form. Two weeks later, the tally in the House was 165, consisting of 97 Democrats and 68 Republicans. In the Senate, 11 Democrats and 4 Republicans had introduced the measure in several variations, all of which embodied its basic thrust. The session of Congress would increase in September with 175 House sponsors, along with the 15 senators.

At that point, the arithmetic looked fine to the promoters of the Bell bill. Usually, even a popular bill gets many fewer sponsors. Ordinarily, 175 supporters, if solid, might be considered enough to ensure passage of a bill.

That, of course, is not the way political dynamics operated. It is, as AT&T management was to learn, much easier to get some congressmen to introduce a bill than it is to get them to vote for—or against—the final version after it has survived subcommittee and committee shakedowns. The calendar is controlled by the leadership, and all congressmen simply were not created equal. Sponsors of CCRA were mainly the rank and file; the leaders were conspicuous by their absence from the list of sponsors.

As the story of the Bell bill and the attempts to rewrite the Communications Act which followed it unfolded, three guiding principles of legislative engineering again became clear. First, it is the kinds of sponsors of a bill, in terms of their position, influence, and dedication, that matter, rather than their sheer number. Second, it is easier to stop a controversial bill from passing, particularly in the Senate, than it is to pass one. Third, mere cosponsorship of a complex bill does not assure continued support once the full impact of the bill becomes known.

Recounted in earlier chapters were efforts to rewrite the Communications Act, concentrated on the broadcast side, and the absence of any really substantive change when it was all over. Now in the wings was the first serious attempt to revamp title II of the act—the part governing common carrier regulation.

The Communications Subcommittee, with Chairman Lionel Van Deerlin (D., Calif.) and Rep. Louis Frey, the ranking Republican, working closely together, decided to hold three days of exploratory hearings at the beginning of October 1976, after the congressional session had adjourned. If the swarm of interested observers had known that the mob scene which followed was typical of many in the coming years, some might have looked immediately for other jobs.

Scenting a major shift in the national policy scene, everyone with even a remote interest in communications either was on hand for the hearings or had a representative there. The Commerce Committee's main hearing room in the Rayburn Building is a good-sized one, and every square inch of available space was jammed by the audience, with many more standing than seated.

After hearing more than forty witnesses during the three-day hearing, and mulling things over for a week, Van Deerlin and Frey announced plans for a basement-to-attic rewrite of the Communications Act. The project focused on title II, the common carrier section, and proposed that changes in other parts of the law always were related to common carrier revisions.

As the new Congress convened, and the early winter months went by, some handwriting began to appear on the wall. Members who wished to have the Bell bills before Congress had to reintroduce them, since all of those offered in the preceding Congress had now expired. By the end of February 1977, the number of House members introducing CCRA—both the original version and one eliminating proposed authority to the FCC to approve mergers—had shrunk from the prior peak's 175 in the house and 15 in the Senate to 42 in the House and 4 in the Senate.

There now appeared a countermeasure, a proposed congressional resolution endorsing competition in telecommunications. At the close of February, it had been introduced by 23 House members and 2 Senators.

Soon thereafter, John deButts, in an interview with *Telecommunications Reports*, commented, "Very few bills come out of Congress exactly as they are proposed." But he added, "The thing that is important to me is that we now have the attention of Congress. They realize that this is a problem."

A problem it was, but the tide was turning. A month later, Rep. Timothy Wirth (D., Colo.), a prominent member of the House subcommittee, stated, "Practically speaking, the Bell bill is dead, and no legislation bearing the name 'consumer communications reform act' is going to be reported by the House Communications Subcommittee during the 95th Congress."

In truth, getting sponsors for CCRA in the new Congress was proving to be a lot tougher job, as the discussion and rhetoric escalated. Near the end of June, CCRA reached a near peak for the 95th Congress with 94 House sponsors and 9 in the Senate. But Tim Wirth was right; CCRA in its original form was dead. Meanwhile, 40 congressman and 6 senators had introduced resolutions calling for congressional resolution of national telecommunications policy. John deButts was right, too; Congress realized that there was a problem.

Earnest and long staff work, discussions among members of Congress, and highly concentrated lobbying went on, as the legislators and their staffs struggled to familiarize themselves with a foreign lexicon, a complex industry structure, and the effects that policy changes already were having and legislative change would produce.

It was not until the middle of the next year, in early June 1978, that Van Deerlin and Frey came forward with a 217-page "Communications Act of 1978." Its hallmark was relatively limited regulation of interstate and foreign communications "to the extent marketplace forces are deficient." In other words, the Bell System's monopoly would continue to be fairly tightly regulated, while the competitors would enjoy some freedom from the hand of government while they worked to make all services competitive in the marketplace. The Western Electric Co. would have been divested.

During the intervening period, leaders of a wide cross-section of telephone companies, Bell and small and large independents, were working on a comprehensive set of recommendations. At the beginning of December 1977, the group transmitted to Congress as a conceptual framework what had to be the most sweeping self-initiated structural change in the more than 100-year history of the telephone industry.

Perhaps the main point underlying the task force's statement was that industry leaders recognized that, however much they may have preferred the traditional industry structure and how much they longed for its return, it was simply impossible to retain it, and sweeping changes must be accommodated.

The industry leaders recommended some basic changes in the traditional regulatory approaches while acknowledging the legitimacy and inevitability of specialized private line systems, private communications systems, and customer-furnished terminal equipment. But they could not accept the concept of open competition in conventional long distance telephone service.

The 95th Congress ended its two-year life without any serious effort to move new legislation. Van Deerlin had described the June 1978 legislative proposal floated by Frey and himself as "only a starting place," and clearly that is what it was. Another Congress was coming, and with it new bills, drafted by legislators in positions of prominence and their staffs, in practice of the old maxim that "politics is the art of the possible."

The Senate had been relatively quiet as work proceeded, mostly behind the scenes, during 1976-78. Now, in March 1979, two key members of the Senate Commerce Communications Subcommittee, each with a few co-sponsors, introduced bills. While tailored to reflect some individual differences of Sen. Ernest F. Hollings (D., S.C.), the subcommittee chairman, and Sen. Barry Goldwater (R., Ariz.), the group's ranking Republican, they were similar in approach.

The bills provided for gradual deregulation of competitive services, retention of the existing regulation of monopoly services at least for a transitional period, broad-scale interconnection of competing interexchange carriers to the telephone network, with access charges to defray the cost of local service, and permission to regulated carriers to engage in other lines of activity, to negate in large part the tariff solution restrictions of the 1956 consent decree.

Soon thereafter, Van Deerlin and two co-sponsors opened the legislative maneuvering on the House side of the 96th Congress with a relatively more sweeping deregulatory bill, for everyone except the Bell System. It would provide generally for another decade of traditional regulation of the Bell System; no regulation at all for the other interexchange carriers; state regulation of basic local service, but none of other intraexchange services or terminal equipment; and state-by-state handling of subsidies for high-cost exchanges. Unlike the prior version, this bill did not contain a provision for divestiture of Western Electric. Van Deerlin described the earlier divestiture directive as a "monstrous obstacle" in the path of new legislation.

It would not serve the time of most readers well to recount every major legislative event of the next several years. Consequently, we will select those that seem most significant. Crises were constant; the congressional path could best be described as a roller coaster. Reporters and observers surveying the situation on a daily basis found it a long series of ups and downs; one day, the pending bill was almost sure to be passed by the current Congress, but the next, it was in serious trouble. On the third day, it was probable that the outcome was totally clouded in doubt.

Culminating a summer of parlimentary maneuvers, shifting and changing alliances, and continued intense lobbying, leaders of the Senate and House committees had some comments for reporters when Congress returned after a brief recess for an early fall session starting in September 1979. They told them that if industry officials did not stop squabbling and take the lead, the legislation could never pass in the current Congress, which had more than a year to go. In prophetic words, Hollings said, "It would be a shame if legislation is impossible to enact," but that he and other committee members "will suffer no personal tragedy if there is no legislation. There are many other things for us to do in the Senate."

As the House labored through 1980 toward a relatively early close in a Presidential election year, what had seemed at one time to be unanimous subcommittee support began to erode. Major disputes were over the access charge arrangements and, more heatedly, over intra-AT&T information flow between the parent organization and separate subsidiaries which would be established. Nevertheless, the subcommittee, for the first time, reported the comprehensive legislation to review title II by 10 to 2. The full Interstate and Foreign Commerce Committee, in another first, then sent the bill on its way to the House floor at the beginning of August by a 34 to 7 margin.

Ordinarily, that is, the bill would have been in line for floor consideration. In this instance, Chairman Peter W. Rodino of the House Judiciary Committee, with jurisdiction over such matters as antitrust litigation, had expressed considerable concern. The Justice Department's suit against AT&T had been pending for four years. Rodino said it was obvious that the bill, if passed, would affect the suit. He proposed eliminating all provisions of the bill dealing with the Bell System structure until the pending suit was finally resolved.

Rodino asked for "sequential referral" of the bill to his committee, amid cries of supporters of the legislation that time was too short as it was. With adjournment drawing near, very hurried Senate action would be needed if the bill was to pass. If Judiciary were to delay matters for long, chances of passage would dim from slim to none.

Cognizant of the political realities, and wishing to keep committee lines of authority clear, House Speaker "Tip" O'Neill agreed to sequential referral but he told Judiciary to report within a month. It did, giving the bill an "adverse" report, but "without prejudice to consideration in the 97th [next] Congress." That took care of matters for the 96th Congress, which adjourned soon thereafter.

The details of the remainder of the story would be fascinating to students of political science, but at the bottom line prove only that the federal government is still the tripartite body which the founding fathers envisioned. Developments in Congress from then on, 1981 through 1984, were sometimes exciting to legislative observers but always keyed closely to events in at least two federal courtrooms, and in the Justice Department and the FCC. Governmentally, the FCC is not technically a part of the Executive Branch, but is a "creature of" and accountable to Congress. However, on telephone policy issues in the early 1980s, it was obvious the FCC was taking its policy-making signals from the Executive Branch.

As will be recounted in the next chapter, at the close of the Carter administration, in late 1980, AT&T and Justice reached agreement on a consent decree. The incoming Reagan administration did not have the right Justice officials in place to ratify it, and Judge Harold H. Greene insisted that the trial go on as scheduled. The Republicans had taken control of the Senate in the 1980 elections, and the new chairman of the Senate Commerce, Science and Transportation Committee, Robert Packwood (R., Ore.), decided that with the consent agreement now off the track, the time had come to proceed with domestic telecommunications legislation.

The Senate stayed on top of the legislation most of 1981, working to a compromise on the toughest issue facing it at the time. This was a provision allowing AT&T to provide electronic information services, bitterly opposed by the potent American Newspaper Publishers Association. By October 7, in what appeared to be a landmark, one house of Congress finally adopted a rewrite of title II. The Senate passed the bill, similar in thrust to the preceding year's House legislation, by a landslide 90 to 4 margin.

Although the bill now had proceeded further than any earlier effort, few were fully satisfied with the Senate measure. With Van Deerlin having gone down to defeat in the last election, new subcommittee Chairman Tim Wirth was having trouble putting together viable legislation in the House. Finally, he introduced what was described as a consensus bill in early December.

What looked like a strong possibility of completed legislation which would be distasteful to the Reagan adminstration and AT&T comprised part of the impetus leading to the early January 1982 announcement of agreement on a consent decree. Ultimately, the decree derailed the bill in the House, with opponents urging that full opportunity be given for the decree to work. Meanwhile, however, the House subcommittee cleared the way for an amended version of the Senate bill in late March, leading AT&T to launch an unprecedented advertising/mail/lobbying program against the measure. Once again, the theme was that the bill would interfere with the consent decree.

In early July 1982, on the eve of a scheduled full committee markup (acting on successive proposed amendments until a final vote was reached), the Reagan administration delivered the fatal blow to the bill. The White House stated its opposition to a bill "that would conflict with the substantive terms of the proposed settlement" of the antitrust case, "or with its effective implementation."

The markup proceeded through more than half of a contentious July, through bitter parliamentary wrangling. Administration supporters insisted that the rules be followed and the entire bill read word by word. Then they offered a lengthy series of amendments. Supporters of the measure noted gloomily that their adversaries could follow the same tactics on the House floor, and that if a joint House/Senate conference report ever could reconcile the differences in the two measures, a Senate filibuster could follow.

At the start of the scheduled third week of full committee markup sessions, Wirth threw in the towel for 1982 and the 97th Congress. He made a five-minute concession speech and moved that the committee adjourn. The bill's supporters, he said, had enough votes to win on the critical issues but did not have enough time to get the bill through the House and Senate.

To a significant degree, history repeated itself in the 98th Congress of 1983-1984. This time, the crucial events taking place off Capitol Hill, described in the next two chapters, were the entry of the consent decree, or modified final judgement; the actual divestiture January 1, 1984, a far more sweeping result than ever considered in Congress; and the FCC's willingness to compromise and delay its access charge program, under strong congressional pressure.

In July 1983, leaders of the two congressional committees with jurisdiction tried a new tack. Announcing their unanimity in "thrust and purpose," although their proposed legislation differed in specifics, the two committee chairmen, Senator Packwood and Rep. John D. Dingell (D. Mich.), announced that they would hold two days of unusual joint hearings. The committees would be looking at bills described by Packwood as "using long distance rates to subsidize rural and residential service," and by Dingell as "maintaining the status quo" (no access charges on local subscribers). Numerous other congressmen were lining up in favor of bills imposing a flat prohibition on end user access fees.

Before standing-room-only crowds in the two largest meeting rooms on Capitol Hill, the Senate and House caucus rooms, a flotilla of witnesses divided sharply on the merits of the bills. Opposition was led by the chief executives of four major interexchange carriers, including AT&T and MCI, top officials of the seven newly created Bell regional holding companies, and FCC Chairman Mark Fowler. On the side of overturning the access program were spokesmen for state commissions, small exchange carriers, and consumer groups.

The day before the joint hearings started, with the congressional drumbeat registering loud and clear, the FCC adjusted its numbers. Now, the residential end user fee would be $2, instead of the former $2 to $4 range. Business users would still pay a flat $6 per line. The premium surcharge imposed on AT&T would rise to $2.2 billion from the previously calculated $1.4 billion.

The Commission's changes made support of the access system more politically palatable, but opponents of end user fees still did not like the looks of things. The House Energy and Commerce Committee, in late October, favorably reported a bill which would, among other things, rule out subscriber line charges on residential customers and businesses with only one line.

Two weeks later, common carrier communications legislation once again had made it through one house of Congress. After six and a half hours of debate, and a steady series of losses by opponents of the legislation on votes running about two to one, the House adopted the "universal telephone service preservation act" by a voice vote.

As divestiture and 1984 dawned, there was noted in *Telecommunications Reports* a "growing sense of inevitability" about passage of some kind of legislation to revise title II to attempt to insure "universal" telephone service. This was because, after late-1983 passage of the House legislation, there remained nearly a year before final congressional adjournment for the Senate to act. When time ran out for the Senate near the end of 1983, Majority Leader Howard H. Baker, Jr., promised that telephone legislation would be the first order of business in 1984.

Consequently, some of the bill's opponents were moving from attempting to block the bill to a form of damage control, to try to make the measure less of a disaster than they viewed the House-passed legislation.

The legislative outlook changed considerably in mid-January, however, when another compromise rabbit came out of the FCC hat. The Commission decided to delay imposition of end user access charges on residential and single-line business subscribers for more than a year, to June 1985. The interexchange carriers would make up the differnce in the settlements "pot" by paying a higher carrier common line charge, which they of course would pass along to long distance telephone users. The competing or other common carriers would pay 45 percent of the premium carrier common line charge, an effective discount of 55 percent for what they perceived as inferior connections.

It did not take an advanced political scientist to be aware that the similarity of the FCC action to a just-received congressional proposal, as well as its timing, was something more than a coincidence. A short time before, 32 Senators, led by Bob Dole (R., Kans.), who later suceeded Baker as majority leader, had written to the FCC proposing almost exactly the action which was taken by the Commission. Senator Dole remarked that he knew personally of another 10 to 15 Senators who would vote against acceptance of the House bill as it stood.

Within a week, the Senate effectively put an end to the legislation for the 98th Congress at least, even though it would not adjourn until the fall. A motion by Senator Goldwater, a critic of divestiture and of the legislation, to table the motion to bring the bill up for Senate consideration was adopted by a 44 to 40 margin.

At the same time, it was obvious that the vote for the motion to table, or set aside, the bill could have been larger had a test really loomed. Four co-signers of the Dole letter voted against the motion, but only after the outcome was assured, for some undescribed political reasons. Several of the letter's co-signers did not vote at all, but it was announced that at least two of them would have gone along with Senator Goldwater had they been present.

Making the best of an apparent loss, Senator Packwood declared victory. He commented, no doubt correctly, that it was the pendency of the legislation aimed at assurance of universal service, and delay and possible reduction of the end user subscriber line charges, which caused the FCC to take the actions it had approved.

Congressional leaders had another opportunity to take a bow late in the year when, as recounted at the conclusion of chapter 15, the residential and single-line business monthly charge was set for mid-1985 not at the previously planned $2, but at $1. The $2 charge did not go into effect until another full year was allowed for transitional purposes.

Some observers, including one of the authors, have publicly expressed the opinion that it is, under present circumstances, just about impossible for Congress to pass comprehensive telecommunications legislation. Main reasons for this view are the tremendous number and diversity of heavily involved interests (the intense difference between the telephone companies and the nation's newspapers concerning the electronic publishing portions of the legislation is a good example of this) and the already described ability of agencies such as the FCC to come up with appropriate compromises if passage of an undesirable bill is seriously threatened.

An answer as to whether this thesis is totally accurate may have to await future events. What is self-evident is that Congress had a great deal of influence on the course of developments, even if it consistently failed to pass any legislation.

It is equally apparent, without too much hindsight, that CCRA was a mistake. "Getting Congress's attention," in 1976, as it was described by John deButts, only resulted in very rapidly putting AT&T on the defensive. It finally forced the company to mount a huge campaign against the House-approved bill in 1983, using a lot of political capital for an essentially negative purpose.

The lesson obviously had quite a bit to do with the AT&T decision to go along with the divestiture consent decree. By then, in view of the congressional reaction, AT&T had no more public policymakers left to address.

18

SEE YOU IN COURT

As a party to an ongoing and long-standing antitrust consent decree, and as the world's largest private corporation, AT&T remained after 1956 under regular and almost routine scrutiny by the Justice Department. This activity was not exactly discouraged by officials of MCI and other competitors who, in the course of their arguments to all available forums that AT&T was engaged in anticompetitive conduct, made Justice staff members regular and frequent contacts.

With all that was going on in the courts and commissions in the early 1970s to bring about a pro-competitive communications environment, it came as no particular surprise when word leaked out in December 1973 that Justice had served CIDs (civil investigative demands) on AT&T. Reportedly, they raised questions which had been asked elsewhere as national policy moved in the direction of wider competition in interexchange communications and equipment supply.

AT&T, with its usual thoroughness, replied with voluminous memoranda and other material spelling out its positions on all of the current issues. During the late summer and early fall of 1974, several meetings were held at Justice with the company. AT&T continued to contend, among other things, that the government had already been down on the antitrust complaint road once, and that the company remained under the tightening strictures of the 1956 consent decree. As the overall business became more competitive, the tariff solution limiting the Bell companies to regulated, tariffed activities was creating further potential difficulties.

When the government slowly returned to regular business after the Watergate trauma, it became clear at the Justice conferences that Attorney General William Saxbe, in the words of one participant, was "quite committed" to filing suit. The subject was discussed at cabinet meetings on two or three occasions, although evidently no formal action was taken. It appeared that the White House was more in a posture of not objecting to the action.

President Gerald Ford was in Japan, on November 20, 1974, when what turned out to be the final blow against nationwide regulated telecommunications monopoly in the United States was filed with the district court in Washington. There is no evidence in the published history of the time that the traveling White House entourage took any unusual note of the occasion.

Some observers have pointed out that the legal counsel to the President at the time was William Seidman, an old and close friend of President Ford who was an Ann Arbor, Mich., attorney. In June of that year, two months before Ford became President and Seidman joined him at the White House, Seidman had taken part in a Michigan Bell Telephone Company rate case. As counsel for the competitive equipment suppliers, the North American Telephone Association, he objected to Michigan Bell's practices regarding customer premises equipment. Saxbe later was counsel to other Ohio competitive equipment suppliers.

It is interesting to compare Justice's request for relief in its fourteen-and-a-half-page filing, of which the first ten and a half pages consisted of background and description of the parties and the final outcome of the suit. Justice once again sought divestiture of Western Electric. Further, Western would be ordered to divest enough in assets, and divide itself into two or more companies, to assure competition in manufacture and sale of telecommunications equipment.

The other main prayer for relief was that AT&T "be required, through divestiture of capital stock interests or other assets, to separate some or all of the Long Lines Department of AT&T from some or all of the Bell operating companies, as may be necessary to insure competition in telecommunications service and telecommunications equipment."

Justice's list of charges of monopolistic and conspiratorial conduct also was somewhat more limited than its testimony later alleged, and again keyed substantially toward Western Electric.

The Bell companies, it was stated, "attempted to obstruct and obstructed" the interconnection of specialized common carriers, miscellaneous common carriers, radio common carriers, domestic satellite carriers, and customer-provided terminal equipment; "refused to sell terminal equipment to subscribers of Bell System telecommunications services"; "caused Western Electric to manufacture substantially all of the telecommunications equipment requirements of the Bell System"; and "caused the Bell System to purchase substantially all of its telecommunications equipment requirements from Western Electric."

Then began a series of events culminating in one of those accidents of history that alter the course of events, and for which no person is responsible. It started with the assignment of the Justice complaint—evidently by the customary rotation system—to District Judge Joseph C. Waddy.

Judge Waddy had been on the bench for twelve years, five as a D.C. Municipal Court judge and seven in the U.S. District Court. He gave every evidence of being a kindly, amiable sort who was genuinely pained when he had to rule against someone. There was no question of his competence, although he had never handled antitrust litigation before. One observer once commented, "If I were on trial for rape, I'd like him to be the judge."

For the next three and a half years, the case went along at a relatively leisurely pace. The tone was set in a few weeks after the suit was filed, when AT&T asked for an extra two months, to February 10, 1975, to respond to the Justice allegations. Noting the complexity of the case, Judge Waddy had no difficulty granting the additional time.

In the early skirmishing—and "early" in this case meant about three years, from 1975 through 1977—AT&T relied heavily on its view that the industry already was pervasively regulated, and that conduct approved, permitted, or directed by appropriate government authority should not be subject to the antitrust laws. The Justice suit, it was argued by AT&T, was *res judicata* (already tried and decided).

Judge Waddy, in his customary style, took his time on the jurisdictional questions and listened to arguments from all sides. In July 1975, he heard four hours of oral argument on the jurisdictional question, after having already received extensive briefs from the two contending parties. Still moving cautiously, the judge then asked the FCC to give him a friend-of-the-court brief on the subject of whether pervasive regulation would bar his court from considering the case.

The Commission, after being allowed several extra months for the filing in the typical fashion of the case up to that point, came up with a response more in agreement with AT&T than not. It told Judge Waddy, "Some means must be found to accommodate the regulatory and antitrust regimes." To the FCC, the answer lay in the doctrine of primary jurisdiction, under which antitrust courts in the past referred issues of fact and law under the Communications Act to the regulatory agency. The courts, as in the Carterfone case, then could apply the antitrust laws as needed in the light of the FCC's findings.

Primary jurisdiction, as between the FCC and his court, obviously troubled Judge Waddy. Having received a maximum of input on the subject, he waited three quarters of a year, until early October 1976, before coming up with a partial ruling. He reached conclusions on two of three gut issues he had posed. *Res judicata* would not apply, the jurist reported. Jurisdiction would not be surrendered to the U.S. District Court in Newark, which still had the 1956 consent decree.

Judge Waddy took considerably longer than the hard-driving Harold H. Greene to reach decisions, but on jurisdiction he came out at about the same place. In mid-November, after listening to still another extended oral argument on the basic FCC/court jurisdictional issue, he ruled that his court was not barred by the "inconsistent standards" of the Communications Act and the antitrust laws from considering conduct alleged to violate the antitrust statutes.

Orally, he ruled that his court had "probable antitrust jurisdiction." Later, in a written order, he chopped out "probable" and declared that "the court is satisfied that it has antitrust jurisdiction of at least some of the aspects" of the case.

Although Judge Waddy's retirement from the bench was a year and a half away, he had a small role in the remainder of the AT&T/Justice case while it was still assigned to him. Most of 1977 was consumed with AT&T appeals to the Supreme Court and the U.S. Court of Appeals for the District of Columbia against Judge Waddy's jurisdictional rulings. They were consistently turned down, and by early December AT&T had exhausted the basic jurisdictional argument, and the parties turned to a new round of procedural arguments.

For a full year, any progress in bringing the case to trial had gone down the tube. Justice had urged the start of the discovery process—the expensive, laborious, slow-moving production of thousands of documents—in February 1977, but it had been delayed for a full year while AT&T's jurisdictional appeals were pending. By December 1977, with the jurisdictional appeals now behind it, Justice once again proposed the start of pretrial discovery, more than three years after the complaint had been filed.

As it happens with many antitrust suits, the process was one of a mountain bringing forth a mouse. One casualty of the slow process was Philip L. Verveer, who headed the initial trial staff preparing for the case. It had been obvious since the filing of the case that the Ford and then the Carter administrations did not consider the AT&T lawsuit the culmination of their missions. Neither administration was in any hurry at all, and neither was ready to put much in the way of money and manpower into the case.

Even those who disagree with Phil Verveer on philosophical grounds would have to describe him as intelligent, conscientious, and otherwise qualified to try an antitrust case against AT&T. Nonetheless, his high-water mark was the victory before Judge Waddy on the jurisdictional arguments. Discouraged about the lack of resources being put into the AT&T case, he resigned as the head of the trial staff. Later, he became chief of the FCC Common Carrier Bureau, before going into private practice.

Judge Waddy's last major decision in the case came in February 1978 and was one which Judge Greene probably would not have agreed with. In fact, Judge Greene issued a directive to the contrary soon after he took over the case. At issue was Justice's proposal that the "fruits of discovery" in seven private antitrust suits and several regulatory cases be admitted in the government case. Since the discovery processes in the private cases, and particularly the major proceedings involving MCI and Litton Systems, were vast undertakings, Justice foresaw tremendous savings in time, money, and personnel.

Judge Waddy, however, denied most of the proposed pretrial orders which had been prepared by the parties for his signature. He said any move to admit the products of discovery from other cases, being tried in other courts, should come from those tribunals. He did not, he said, believe he should "inject myself" into the other cases.

At that point, start of the trial appeared far in the future. Under the very best of circumstances, discovery in a case of this magnitude is likely to take several years. In this situation, the parties were not even operating under orders to get the trial preparation fully under way.

A U.S. magistrate, Lawrence S. Margolis, handled the next status call. He ruled after hearing argument that AT&T turn over the products of discovery in ten judicial and regulatory cases as sought by Justice. AT&T promptly appealed the ruling to Judge Waddy.

Judge Waddy never had an opportunity to act on the appeal. In late June, all of his cases were reassigned to a newly appointed district judge, Harold H. Greene. Judge Greene had headed the D.C. Superior Court, and he was described at the time, in a triumph of understatement, as a jurist "who keeps the cases before him moving." He immediately directed counsel in the AT&T/Justice case to file "succinct status memoranda" within twenty days.

In little more than a month after his cases were reassigned, on August 1, 1978, Judge Waddy passed away. With him went the possibility of a much more leisurely proceeding than the one presided over by Judge Greene. It is highly likely, in fact, that the outcome would have changed completely if Judge Waddy had lived and retained jurisdiction of the case.

In light of what happened later, the case almost surely would have ended in a negotiated settlement. The accord would have, it seems clear, been one that was really negotiated, a mutually acceptable compromise, rather than what actually happened—AT&T's acceptance of Justice's terms. In turn, the slower timetable and the less draconian solution might well have caused Congress to adopt comprehensive legislation, and the FCC to act differently than it did. Nonetheless, were it not for the carcinogen which afflicted Joseph Waddy, the Bell System—somewhat curtailed in size and activity by an antitrust settlement with the Carter administration and more closely restricted by congressional enactment and FCC order—probably would still exist today.

Judge Greene, from the beginning, wasted little time. In mid-September 1978, he asserted jurisdiction over the entire complaint, eliminating the FCC's suggestion of a primary jurisdiction referral. He told AT&T to furnish to Justice the 2.5 million documents it had furnished in the MCI and Litton cases. Just to make the rules of the road even clearer, he ordered that discovery be completed in about a year and a half, an astonishingly short time for a case of this size and complexity. He ordered a series of statements of contentions and proof by the parties, spelling out the evidence they intended to rely on in the trial, with a view to progressively narrowing the issues.

The Greene court continued in complete commmand of the litigation, while attorneys and support personnel for both sides labored almost literally around the clock. Justice's first statement of contentions and proof totaled 629 pages, and several of the succeeding statements of the two parties easily topped that figure. Teams of lawyers and their accompanying experts worked intensively on assigned portions of the mammoth job of discovery. The Supreme Court added another 2.5 million documents to the incredibly huge stack of papers by declining to review Judge Greene's order requiring production of the papers from the MCI and Litton cases.

Almost exactly one year after he was assigned the case, in late June 1979, Judge Greene took another step demonstrating that he would, in fact, "move the case along." He issued a detailed step-by-step timetable, aimed at achieving maximum preparation for a relatively less time-consuming trial that he scheduled to begin September 1, 1980. Taking a good look at the size and complexity of the task at hand, an amazingly short time had elapsed from the date Judge Greene was assigned the case before the pace began to accelerate sharply.

Throughout the pretrial processes, and during the trial itself, Judge Greene's approach became clear. He would keep the parties' "feet to the fire" with difficult time schedules and deadlines, albeit ones he did not consider impossible. If the parties came back and reported that, despite their earnest best efforts, they could not quite make an ordered deadline, he would allow an extension—but it would be the briefest possible.

During the whole procedure, Judge Greene's time strictures were not met to the letter, but the outcome was close—far closer than ever might have been imagined. He advised all concerned that one of his objectives was to show that such huge antitrust trials could be handled in the federal judicial system efficiently and within a reasonable time. There was little doubt that he achieved that goal.

The new, and to Judge Greene absolutely final, date for the start of the trial became January 15, 1981, near enough to the originally slated September

1, 1980, to prove the judge's point. After opening statements, it would be up to Justice to present its entire case, presumably without any recesses of any consequence. The nearly exhausted Justice trial staff had serious doubts that it would be ready. More than two years of extensive work on discovery, stipulations, statements of position and issues, and pretrial briefs had taken its toll.

AT&T had far larger resources, but its management had a business to run while dealing concurrently with the hazards of simultaneous antitrust suit preparation, congressional legislation, and progressively competitive regulatory policies. The resulting uncertainites had taken as big a toll on its top management as the exhausting pretrial work had exacted from the Justice staff.

Both were ready, in the dying days of the Carter administration, to strive for a settlement to head off the trial. Soon before the scheduled trial start, and the inauguration of Ronald Reagan as President, they reached a concept or framework of a settlement agreement. There was, however, not nearly enough time to agree on details and put them into the formal language of a consent decree.

Among other things, the framework provided for the long-missing yardstick to permit analysts to measure the effectiveness of regulation as a surrogate for competition. This would come about by divestiture of three "associated Bell companies"—the two relatively small companies in which AT&T owned substantial minority interests, Southern New England (Connecticut) Telephone and Cincinnati and Suburban Bell, and one in which it owned a substantial majority, Pacific Telephone. Western Electric would be broken into two parts, one of which would be spun off as a Western clone independent of the Bell System.

The concept also called for structural separation of the AT&T Long Lines Department, provider of AT&T's regulated interstate and foreign communications services. This, it was felt, would comply with the mandate of the 1956 consent decree, limiting the company's communications operations to tariffed services. Separating out the provision of unregulated terminal equipment and enhanced services would also be in accord with the FCC's computer II decision.

Judge Greene quickly turned down the parties' request for a postponement of the trial to allow time to develop a consent agreement. Maintaining his consistent approach, he issued a memorandum: "The way to conclude litigation and to dispose of cases is to begin. . . . Once a trial has begun, the disputes between the parties may either be decided on their merits at the conclusion of the proceedings or they may be mooted by a settlement. Such a settlement, in turn, may well be more readily induced by the continuing pressure of the trial than by its indefinte postponement."

Once the trial had started, with a day and a half of opening statements, it was a slightly different story. Still a true believer in the "feet to the fire" approach, Judge Greene told the parties they could have a little more than two weeks' recess, to February 2, to get statements from their principals—including a new attorney general or his designee—that they supported the settlement framework and believed it could be translated into a decree within thirty days thereafter.

The statement could not be obtained, either by the original February 2 deadline or within the next month. Attorney General William French Smith, who had served as a Pacific Telephone director, and Deputy Attorney General Edward Schmults, who had been a legal counsel to Bell companies, recused themselves where matters involving AT&T were concerned. By Judge Greene's extended deadline, the man who would serve as Mr. Justice Department where AT&T issues were concerned, Assistant Attorney General-antitrust William F. Baxter, had been nominated but not confirmed by the Senate and sworn into office.

Judge Greene had waited long enough, and testimony began on a four-day-a-week basis. That gave the hard-pressed lawyers and support people on both sides weekends, evenings, and nights, and usually one business day a week, to complete the huge stacks of testimony, stipulations, and exhibits which filled the long days in court. Some observers expressed the belief that the only reason the trial ran four days a week usually, and not five, was not for their benefit but to give the judge a day a week to clear up the other matters which had been assigned him.

The early-1981 effort to settle the case with a consent decree viewed compatible to computer II having failed, AT&T, finding itself hung up in potential conflict between the 1956 consent decree and the FCC order, went to another court. It asked the U.S. District Court in Newark, N.J., which had retained jurisdiction over the 1956 decree, to rule that the agreement did not prevent the Bell companies from furnishing or manufacturing detariffed customer premises equipment (CPE) and providing enhanced services, also not under tariff. CPE and enhanced services, it argued, remained "subject to public regulation" even if no longer subject to tariff filing requirements.

Justice opposed the AT&T request, as did various competitors in the CPE and enhanced service markets. The case was assigned to Judge Vincent P. Biunno, who held an almost immediate all-day oral argument. The judge seemingly "telegraphed" his decision, issued four and a half months later, making comments which appeared generally sympathetic with AT&T's position that the drastically changed circumstances of the quarter century since the consent decree required a new look at the decree's requirement that, in general, it provide equipment and services "subject to regulation."

In early September 1981, Judge Biunno issued the first major construction of the decree since its entry in 1956. He concluded, "The language of the judgment is clear and unambiguous, and it seems to the court beyond dispute that AT&T, in complying with the FCC order, will be engaging in the business of furnishing communications services and facilities, the charges for which are subject to public regulation under the Communications Act of 1934. The judgment here does not stand in the way of implementing the order.''

Intensified activity regarding Bell System structure and functions was going on at every major level of government. A cabinet council on commerce and related matters headed by Commerce Secretary Malcolm Baldridge was working toward a unified administration position on telecommunications legislation and the antitrust suit, but Assistant Attorney General Baxter, although in the third level of the Justice hierarchy, for matters except those related to AT&T, usually was able to maintain control of the situation.

Articulate, well-versed in both theoretical and practical aspects of antitrust law, and seemingly gifted with total recall, Bill Baxter lost little time in making his presence felt on the telecommunications policy scene. He was at his best in the give-and-take of a congressional hearing or a press conference and reportedly did as well in private meetings. In one memorable Capitol Hill performance, he took a bulging briefcase with him to the witness table, fielded complex and often hostile questions readily for several hours without giving an inch, and never opened the briefcase.

Less than a month after taking office, Baxter put to rest any lingering belief that there might be an early consent settlement of the AT&T case. At his first press conference—such sessions were always very well attended, since AT&T matters had peaked among business news as a result of the legislation and the antitrust suit—he reported he had turned down a Defense Department request that the suit be dismissed. Defense took a consistent position on the subject, with the claim that the AT&T network "is the most important communications net we have to serve our strategic systems within the U.S."

The trial would go forward, Baxter vowed, although in the final analysis Defense's views would be seriously considered. But, in the present context, he said in a widely quoted statement, he intended to "litigate the case to the eyeballs."

To a reporter who asked if he would have accepted the Carter administration's proposed consent decree, he came close to spelling out the final outcome of the lawsuit. The decree would have been unacceptable to him, he explained, because it did not separate all of the regulated activities of the Bell System (all of the local exchange carriers) from the competitive enterprises.

Baxter continued, "We are well into the morn of the day when Long Lines should not be regulated at all, and should be put over on the competitive pile.

The [local companies] would be on the other [monopoly] pile." The competitive pile, he continued, would include, as well as Long Lines, the selling of telephone equipment, information and data processing services, and manufacturing.

In a little over four months, Justice presented ninety-three witnesses to wind up its direct case in early July. None of the testimony produced any dramatic surprises and typically covered events—many of them recounted or referred to in preceding chapters of this book, along with internal AT&T minutes and memoranda dealing with them—which were well known to the parties. The sharp differences between the positions of Justice and its witnesses, and those of AT&T counsel as indicated in cross-examination and objections, related to interpretations and analyses of the developments, rather than the facts brought out.

The conclusion of the plaintiff's direct case triggered two moves which appeared to be landmarks, although the long-run impact of the first probably was conjectural. Bill Baxter, making it clear for the first time that he was speaking not just for the antitrust division or Justice, "but for the administration and the President," asked Judge Greene to suspend the antitrust trial for nearly a year, to June 30, 1982, to remove an obstacle to congressional consideration of the pending legislation to amend the Communications Act.

Baxter explained in a hearing in Judge Greene's chambers, the record of which was quickly made public by the judge, that the administration had concluded, "There is no realistic possibility of moving the legislation, which is now usually known as S 898, a very comprehensive deregulation of the telecommunications industry, through Congress unless, in some sense, this case is put on ice." Until the prior week, he went on, he had not viewed the bill as "an adequate substitute for the relief we were seeking in the case."

Now, he went on, a proposed amendment to the bill already reported out by the Senate Commerce Committee caused him to change his mind. The amendment, drafted by the Justice official and subsequently known as "Baxter I," would put a limit on how much equipment Western Electric could sell to the Bell System companies, with the amount to be determined by how much it sold outside of Bell to independent telephone companies and other customers.

Judge Greene, quickly denying the request, ordered that the trial resume as scheduled, August 3, for the start of AT&T's case. He said the proceeding would be terminated if legislation were enacted to moot the court case, and that any consent decree meeting Clayton Act standards would be entered and enforced. But, he declared, "It would be inappropriate for the court to suspend this lawsuit, which has been pending for seven years, in the middle of a trial which is now scheduled to end by December of this year, simply because suspension may have a political impact in other forums."

Baxter's change of heart gave the pending Senate legislation a seeming shot in the arm, especially when he testified at a Senate Judiciary Committee hearing the following week. He said he would be "content, even enthusiastic" about the bill if it contained the "fair market" equipment amendment he had described to Judge Greene, plus another amendment, as yet undrafted, to give network access "of equal quality at equal prices" to AT&T's long distance competitors. The equal access amendment later proposed, usually known as "Baxter 2," bore a strong similarity to the equal access provisions ultimately agreed on in the consent decree.

As we have seen, the legislative way to resolve Justice's concerns was simply not in the cards. Back in court, Judge Greene opened the way for the eventual draconian solution a short time later.

At the close of Justice's direct case, AT&T and its attorneys faced a choice. It would have been customary in such lawsuits for trial counsel George Saunders to submit a brief written or oral motion asking dismissal of the Justice complaint because it had not proved its case. It would have been equally usual for Judge Greene to summarily deny the motion without further comment and order the defendants to go ahead and put on their case when the trial resumed in slightly less than a month.

AT&T's second option was more in keeping with the company's thoroughness and dedication to its long-held belief in itself and its perceived mission. It was the course that George Saunders and those who paid his fees decided on. AT&T, with the conclusion of the government's case in early July 1981, filed a 553-page stack of pleadings and documents asking dismissal of the Justice complaint "in its entirety."

The size and nature of the dismissal request challenged the judge to respond at some length; a one-sentence denial would have been within his authority, but not within his nature as an active, involved presiding jurist. His seventy-four-page opinion released September 11, 1981, gave his views in some detail, and for all practical purposes brought the antitrust case and the integrated Bell System to an end.

Evidence presented by the government, Judge Greene stated, gave him no reason to dismiss the complaint. Instead, he concluded, it was now up to AT&T to try to rebut the charges in all except three comparatively minor areas. He held, "The testimony and documentary evidence adduced by the government demonstrate that the Bell System has violated the antitrust laws in a number of ways over a lengthy period of time."

His real crusher was this: "Of the three principal factual issues—whether there has been proof of anticompetitive conduct with respect to the interconnection of customer-owned terminal equipment, the Bell System's treatment of competitors in the intercity services area, and its procurement of equipment—the evidence sustains the government's basic contentions, and the burden is on the defendants to refute the factual showings made in the government's case in chief.''

As we have noted, there had been little dispute about the factual evidence all along, and refuting it would have been a virtual impossibility. In theory at least, AT&T could still have persuaded Judge Greene to draw a conclusion from the facts more favorable to it, but to do so meant an uphill battle against nearly insurmountable odds.

AT&T and its lawyers tried valiantly. They produced many more witnesses than Justice and had scheduled a lengthy series of further appearances before the consent decree brought the trial to an end. Present and former company officials earnestly testified as to their understanding of past events, what the companies had done, and why they had done them. Former regulators and other government officials decribed their recollection of official actions and intentions, including the strengths and infirmities of pervasive regulation.

Several witnesses from outside Bell ranks synthesized much of the company's closing case. The 1972 Democratic Presidential candidate, former Senator George McGovern, was concerned about "unraveling the principle of the regulated monopoly." Dr. Harold Brown, former Secretary of Defense, warned that breaking up the Bell System would erode national defense capabilities. Dr. Eugene V. Rostow, who was currently director of the Arms Control and Disarmament Agency, and who headed President Johnson's task force on national communications policy, said the latter group found that the switched telephone network "was indeed a natural monopoly."

Judge Greene was obviously growing increasingly restive as the AT&T case went on. He sharply criticized the length and makeup of AT&T's proposed January witness list, making it clear he did not intend to listen to a lot of prominent people if their testimony was not directly relevant to the trial. After hearing extensive testimony by AT&T witnesses on various competitive responses the company had undertaken, he chilled company management with the observation that he viewed competition by AT&T only as "taking something away" from new entrants.

More and more, it was looking as if the judge would find the company guilty in several key areas, and that only the precise findings—and probably more importantly, the relief to be imposed—remained in doubt.

There was, in early November, another signpost on the road to divestiture for the alert observer. It came in another congressional hearing, during the time when the Senate had passed S 898 but the House Energy and Commerce Committee, the one with legislative jurisdiction over the subject, was still trying to get its act together. The unlikely locale was the House Government Operations Information Subcommittee, which had no authority to initiate legislation but whose chairman had discovered there was a lot of interest, and some publicity, in communications topics.

On one side of the table was the cornpone chairman, Rep. Glenn English (D., Okla.), who reported "confusion" because "I don't see what in the world this little old bill [S 898] has to do with that big old lawsuit." On the other side was Bill Baxter, the urbane, intelligent antitrust expert who carefully refrained from his usual chain smoking while in the witness chair, but obviously detested the necessity for such unproductive sessions.

Trying to explain the relationship of the "little old bill" and the "big old lawsuit," Baxter stressed that the relief being sought by Justice—separation of the local exchange monopoly companies from the remainder of the Bell System—would meet the structural objectives of the legislation. He added, "We don't care much if the local operations are transferred to a single company or three or five or 23 separate companies. The remaining activities of AT&T are competitive."

The trial before Judge Greene went into a short holiday period, with its presiding jurist insisting it would finish in January. Secret negotiating sessions between AT&T and Justice personnel went on at a frantic pace, and they were among the few in the capital working long and hectic hours as the year's end neared. Satisfied that things were on track, Baxter departed for a Colorado skiing vacation.

On December 31, 1981, a statement was issued in Baxter's name that the antitrust division had entered into settlement talks with AT&T "which, it is hoped, may lead to the settlement of pending litigation."

19

THE WORLD CHANGES

The responsibilities of the chief executive officer of AT&T in the many years of the integrated Bell System were awesome. Top management committees and councils existed, but the word of the CEO—the president for many years, and the chairman starting in the 1960s—was final. He set the system's course, selected his own successor, and there is no reported occasion when he was flatly reversed by the company's board of directors. Issues were discussed with the board, and members' views solicited, but the consensus which evolved always was led by the CEO.

That situation is not uncommon in American corporations, large and small. The difference in AT&T's circumstances was the company's size and pervasiveness in the U.S. economy. As suggested earlier in this book, pre-divestiture AT&T was the world's largest corporation by most standards. Its output was an increasingly significant part of the gross national product. It dominated an essential industry whose services and facilities reached every hamlet and every resident.

No CEO was a stauncher defender and guardian of the Bell System's mission, purposes, structure, and role in U.S. society than Charlie Brown. He had not embarked on his tenure as AT&T chairman to preside over the Bell System's dissolution. At every regulatory, legislative, judicial, and political turn in his years as CEO—as things seemed always to deteriorate for AT&T—he remained steadfast in defending its vertically integrated structure and the part it was playing in the telecommunications revolution.

A fairly recent popular book on divestiture, in what seems to be a bit of revisionist history, implies that divestiture of all the local operating companies was a long-planned strategy. A conscious decision was made early, it is indicated, to move into the competitive Information Age by retaining manufacturing and research capabilities and shedding the regulated (monopoly) local companies.

True, that was the result, once the die was cast and there was no other direction to take. It is equally obvious that divestiture of the local companies was one of the options under study for a long time—not to consider all the possibilities would have been irresponsible.

But we do not believe that the top AT&T management decided on this course well in advance, and then adroitly maneuvered its way through the settlement negotiations to make it look as though divestiture was forced on the company, as has been suggested.

What did happen was that at an early stage of his tenure as chairman, Brown realized that management could not indefinitely run the business and carry out its reponsibilities to the shareowners under the overhanging clouds of the government suit and the myriad existing and potential private antitrust litigation. The antitrust proceedings were sapping management's energies and attention and virtually paralyzing decision-making and planning.

In the face of these harsh realities, AT&T did in fact wish an early settlement of the government's challenge to its existing and future structure. The outlook deteriorated even further for the company in the wake of Judge Greene's favorable assessment of the government's evidence and his preliminary finding that AT&T was guilty as charged on most counts.

When Brown took over the Bell System helm, there were many in the company hierarchy who urged him to stay the course, through the trial and the lengthy appellate process which would inevitably follow. Others counseled an early settlement, although none was advocating the result that ultimately came about.

The chairman had some question about the realism of judgment of the inner circle of in-house lawyers who had devoted most of their working careers to the Bell System as it had always been constituted. In the normal course, the retiring general counsel, Mark Garlinghouse, would have recommended his own successor. Brown decided that a fresh approach was required and that he would bring in an outsider as general counsel.

His selection, Howard Trienens, was not an outsider by the usual standards. If not a member of the immediate family, he was certainly a blood relative. He was a senior partner in the big Chicago-based law firm of Sidley & Austin, which for many years had been AT&T's principal outside counsel, particularly in governmental and regulatory matters. Another senior partner was former FCC Chairman Newton Minow. Trienens himself had been in the thick of the most important regulatory and policy-making cases at the FCC. Sidley & Austin provided both strategy and principal counsel in the conduct of AT&T's defense of the government suit and most of the major private cases, including the huge MCI case.

Trienens became a principal spokesman for AT&T in hearings before various congressional committees on telecommunications structure and policy. He was the company's principal negotiator with the Justice Department on possible settlement and, more than anyone else, shaped the final result which so shocked the industry, within and outside of the Bell System.

But it was Charlie Brown who had to make the final decisions for AT&T. His character and nature belie any suggestion that divestiture was AT&T's own idea and make it inconceivable that he would cynically sell the Bell System down the river. A second reason to disbelieve the "they planned it that way all along" theory is the way divestiture was conducted and the results that were achieved for those on whose behalf AT&T was managed—its owners.

In an age of leveraged buyouts, insider trading, "golden parachutes," junk bonds, and incredible management benefits, with wheeler-dealers and management in the saddle and the ordinary common shareowners usually treated as second-class citizens, the spin-off of the local companies was conducted with absolute fairness to the stockholders who would subsequently own parts of all the companies. Extremely high on the AT&T management's priority list was its fiduciary responsibility to the shareowners, and it was not forgotten.

As 1982 dawned, it was becoming apparent that the management could no longer fulfill that fiduciary responsibility by giving the company the best direction it was capable of. The uncertainties were too overwhelming. To finance, build, and operate a pervasive nationwide system, with 1 million employees, is not something that can be handled as a seat-of-the-pants operation.

Beset on every side, facing an imminent adverse decision by Judge Greene, with Congress ready to move if the court ruling was not tough enough, with competitors at all levels of its business cutting down its market share and likely at the same time to take private litigation advantage of a "guilty" verdict by Judge Greene, and looking at a national policy of increased competition by others but with little room of its own to respond, AT&T was fast losing the ability to manage the business.

Immediately after New Year's Day, Washington news people kept a close eye on the scene. The Justice announcement on New Year's Eve that settlement talks were proceeding was on the top of reporters' desks. Between the legislative drive on Capitol Hill and the antitrust suit, the telephone industry had gone suddenly from a little understood and largely ignored source of news to front-page attention among business news topics.

When, late in the morning of January 9, 1982, Justice and AT&T representatives each took a list of Washington news offices and bureaus and began calling to alert them to a special press conference that day, they very quickly had the reporters' attention. The scent of a settlement was big news. The news media people making the calls did not tell anyone what was going on, and they did not have to. The stock exchanges closed trading in AT&T stock.

A big auditorium was needed to hold the crowd of eager reporters, and the Justice/AT&T press people had obtained the largest one in the downtown Washington area—the ballroom of the old National Press Club, larger than anything now available there since the comfortable old club has been renovated in garish modern. Advance press material was distributed, and by the time Charlie Brown, Bill Baxter, and Howard Trienens walked into the room, the reporters had had time to read enough to know the case had been settled.

Several documents were distributed, including a thirteen-page modified final judgment, providing that the 1956 consent decree "is hereby vacated in its entirety and replaced" by the new settlement. Another was a three-line "stipulation for voluntary dismissal" filed by AT&T and Justice with Judge Greene. Baxter gave an unequivocal "yes" to a reporter's direct question: "Does this mean the suit is effectively dismissed, without further action by Judge Greene?"

The proposed modified final judgment was filed with the federal court in Newark which had jurisdiction over the earlier decree, and the parties moved to have the proceeding transferred to Washington and Judge Greene. The two parties expressed the legal opinion—one not shared by Judge Greene, it turned out—that the Tunney Act, which provides among other things for comments by interested parties to the court before a consent decree is accepted, did not apply in this situation. Baxter said, however, "We have proceeded according to the spirit and intent of the Tunney Act."

A reporter asked Baxter how much of the government's goal had been achieved, and Brown broke in, with a rueful smile but obviously not joking, to comment, "It's exactly what the government wanted." Baxter was asked later to comment on the AT&T chairman's statement, and he did not contradict it, remarking that the agreement "meets basic antitrust objectives" and was "highly satisfactory."

Brown was asked, of course, why AT&T had agreed. As he has done many times since them, he pointed to the long-standing and complex policy dispute in which AT&T was enmeshed, and the substantial uncertainties that resulted. He expressed the company's view that the "real essentials" had been preserved for AT&T's customers, shareowners, and employees, and that the outcome appeared to be in the public interest.

A prepared statement issued by AT&T in Brown's name concluded, "In short, we are acknowledging what has already been decided—not in the courts, but through the process by which public policy is customarily made in this country."

In broad outline, the agreement called for complete separation between those services and activities which had become competitive, such as interexchange communications and customer premises equipment, and those which had retained a monopoly or "bottleneck" character, such as local exchange and subscriber access to the long distance networks. AT&T would take the competitve markets. The individual Bell operating companies would retain the monopolies.

Obviously crucial to the whole consent agreement was a five-page appendix to the modified final judgment entitled "Phased-in BOC (Bell operating company) provision of equal exchange access." It bore strong resemblance to the Baxter 2 amendment under consideration on Capitol Hill. Its provisions for equal access were closely in line with what actually happened when more than two-thirds of the nation's telephones were converted to equal access, and their subscribers selected their preferred long distance carrier, in the two years starting in September 1984.

The appendix on equal access provided, "Each tariff for equal access shall be filed on an unbundled basis specifying each type of service, element by element, and no tariff shall require an interexchange carrier to pay for types of exchange access that it does not utilize. The charges for each type of exchange access shall be cost-justified, and any differences in charges to carriers shall be cost-justified and on the basis of differences in services provided."

Within a week, Judge Biunno issued an order in the Newark court transferring jurisdiction over the 1956 consent decree to the tribunal in Washington. Judge Greene announced that he would apply the procedures spelled out in the Tunney Act. He set deadlines for the competitive impact statement required from the Justice Department in that law and for comments on the statement by interested parties.

Charlie Brown quickly became one of the most sought-after public speakers in the United States. A little more than a month after announcement of the consent decree, he faced one of his more apprehensive and somewhat hostile audiences. The state regulatory commissions' association, the NARUC, was holding mid-winter meetings in Washington, and its Executive Committee invited him to address a luncheon session.

Little was known of divestiture's specifics at that point, and the state commissioners were among the most concerned about its ultimate provisions. Many so-called analysts immediately concluded that since AT&T had negotiated the decree, and not the Bell operating companies (BOCs), it was saving the best parts of the business for itself. Readers of financial reports learned after divestiture that this was clearly not true (stock market price trends indicated strongly to the contrary), but at the time it was a popular misconception.

As the public officials who would have to approve rate increases to enable the BOCs to maintain service if their financial viability was in danger, the state commissioners were quick to sound the alarm. Once again, they thought they saw their local customers getting the shaft while the long distance users served by AT&T were getting the gravy in the form of rate reductions, and the FCC would be getting the credit.

Consequently, the main argument of NARUC officials and other state spokesmen during this period was that the BOCs should be divested with all assets intact. This would be subject to a later negotiated split at arm's length between AT&T and the BOCs, once they had established their identity as independent companies.

The states' worries were not assuaged by Baxter's insistence on strict construction of the basic bottleneck theory of the decree: that the local exchanges could provide only regulated monopoly services to prevent their discrimination against others by the use of their bottleneck facilities. The original decree was interpreted, until later changed by Judge Greene, to leave the familiar yellow pages classified business directories in AT&T's hands. It also gave ownership of all customer premises equipment, now well on the way to being detariffed and deregulated by the FCC, to AT&T.

Thus, in darker moments, the states visualized the newly independent BOCs as nothing but transmission belts between the profitable parts of the business. Although they did not regulate yellow pages directories, some states considered the BOCs' directory profits as an offset when fixing telephone rates. Customer premises equipment (CPE) had always been included in the rate base as part of the familiar end-to-end service for which users paid. It not only generated local revenues but also was figured into the substantial and growing amounts interstate long distance service paid for the use of local plant.

In his talk in a large room crowded with state commission officials, Brown first defended AT&T's action in accepting the decree itself, a move which had engendered far from enthusiastic state response. He said "no one could mistake" the consensus regarding a competitive environment intitiated by the FCC and endorsed in legislative debate, and the inhibitions put on the Bell System by the suspicions regarding its size and perceived ability to cross-subsidize. The inevitability of the decree, he told his audience, was simply a fact of life.

Further, he gave little comfort to the states for acceptance of their main contention—that the BOCs should be spun off with all assets intact. The proposal was "entirely contrary to the theory of the decree," he pointed out. It was the AT&T stockholders who owned all Bell System assets, the company's chairman noted, and they were the ones who would receive ownership of the assets in the form of stock in the independent companies to be formed.

The vast impact of divestiture on the American society and economy became obvious once again when public comments were filed with Judge Greene in mid-April. Literally hundreds of statements from competitors, regulators, users, and concerned individuals deluged the court. Almost all of the commenters sought some revision in the modified final judgment.

The Tunney Act filings contained one chilling threat for those who looked to the decree to provide, at long last, some certainty about the future structure of the business. The top legal officers of twenty-four states, in a joint statement, warned of endless litigation on the local level. They said if the state utility commissions decided that the decree "compromises their ability to carry out their important regulatory responsibilities," they might "exercise their power to prevent divestiture," or seek court injunctions, within their states. The post-divestiture period has been litigious enough, but that was one prospect that did not eventuate.

Even though the two litigating parties had agreed on terms for an armistice, Judge Greene remained in charge. Until August, when he issued his order, debate went on as to the scope of his authority. Baxter, for one, contended that the court did not have the power to order changes in the modified final judgment, but others doubted that Judge Greene's hands had been tied by the consent decree.

Characteristically, the judge found a way to require changes he believed were in the public interest without stretching his legal authority beyond its limits. He described the modifications he would require and neatly sidestepped the jurisdictional question by advising the parties that if they did not accept the changes, he would not enter the decree, and the trial would resume.

As observed in *Telecommunications Reports*, "Assuming that the conditions he has placed on acceptance of the Justice Department/AT&T antitrust settlement are accepted by the two parties—as most knowledgeable observers believe they will—Judge Greene may have succeeded in writing a new national telecommunications policy where Congress has failed for the better part of a decade."

The court accepted the basic thrust of the modified final judgment. It ruled that the BOCs must be able to sell new customer premises equipment, while accepting the inevitable premise of the decree that AT&T had to be the owner of in-service or "embedded" CPE. Allowing the BOCs into the competitive and unregulated, but local, communications business of purveying new terminal equipment was one of several measures designated to help assure their financial viability.

Another change was the requirement that distribution of printed yellow pages directories, a lucrative operation, would remain in the hands of the BOCs. Judge Greene made it clear that he had heard the laments of the American Newspaper Publishers Association that everything dear to the nation, starting with the First Amendment, would go down the drain if the telephone companies could produce what amounted to electronic classified advertisements in competition with the newspapers. (Newspaper publishers had nightmares about constantly updated yellow pages containing "today's specials" and "for sale tomorrow only.") The decree retained the ban on BOCs' provision of electronic information, and AT&T was banned from electronic publishing of its telephone customer base for seven years.

Judge Greene left the basic equal access provisions unchanged, other than specifying that, during the interim before equal access conversion, AT&T competitors getting less-than-equal access would pay less for it.

The two parties, to no one's surprise, found that the court had left them with no choice except to accept Judge Greene's suggested changes. They quickly acquiesced, and the decree was entered by the judge before the end of August.

Thus began an incredible period, although in many respects it had started in January with the announcement of the consent decree. The Bell System could not wait for entry of the modified final judgment; there was just not enough time. Committees and task forces had been formed and were hard at work on a bewildering array of assignments which had never been performed before.

The integration of the Bell System made the job of its separation and divorce literally equivalent to unscrambling eggs. The many aspects of the process are mind-boggling, and at least one other book has detailed them; that will not be repeated here. Let us, instead, strive for a general concept of the task at hand.

The electronic system which comprised an unparalleled nationwide network, reaching everywhere and providing in effect the world's largest computer system, had been built without much regard to ownership other than Bell System. There were often incomplete inventories of property which at least provided the financial records necessary to divide the plant between interstate and intrastate, as required for cost separations. Technically, but sometimes not in fact, there were records which delineated ownership between AT&T and the BOC serving the area and could be used for asset distribution. Remember, however, that arbitrarily assigning ownership would not be enough. The combined operation, now divisible into its separate parts, would function under split ownership, but it would still have to keep operating just the same as it did when everything was owned by the AT&T shareowners.

A huge problem involved real estate. The obvious and efficient pre-divestiture mode was for AT&T to occupy portions of BOC-owned buildings and vice versa. After all, they were all doing the same thing for the same customers and the same employer. Now, one or the other had to own the real estate. The simply solution, put in effect, was that each location was owned by the company which had predominated in its pre-divestiture use.

The other, AT&T or the BOC, would be a tenant, to avoid disrupting things. Fine, if there is plenty of time to work things out; but as of 12:01 a.m., January 1, 1984, the vigilant regulators and perhaps the attorneys for angry customers, competitors, or stockholders had to be convinced that neither party had been treated unfairly.

An entire book could be written on the financing complexities which resulted. Remember that, with minor exceptions which were necessarily but not easily wiped out in the two years between the announcement of the decree and the actual divestiture, AT&T stockholders owned all of the equity (non-debt) of the Bell System. They were the owners, with all that entailed, of large companies which looked liked substantial organizations in their own right: New York Telephone, Illinois Bell, Southern Bell, and twenty more like them.

Financing operations of the pre-divestiture Bell System appeared to outsiders to function like a finely tuned Stradivarius. Most of the debt capital, raised under neat schedules keyed to the bond market and the concerns of underwriters and buyers, was issued in the names of the BOCs. After all, they were the ones with physical property that could be mortgaged. When the market required, AT&T sometimes sold large issues of debt. But these were usually debentures, secured by the general credit of the company rather than a mortgage on physical property, and they were normally convertible into common stock to become shares of ownership, rather than "chits" which had to be paid off within a specific period. Most of the Bell System's equity capital was raised by AT&T, which made advances to the subsidiaries. The latter then would repay them by issuing additional capital stock to AT&T.

On the equity side, the principle was fairly simple. Once the BOCs were divided into independent companies, shares of ownership would be "spun off" to the AT&T stockholders in some established manner. Debt was something else again. With the exception of Pacific Telephone, the debt ratios of the BOCs were more or less the same. In this situation, more-or-less was not good enough. When the Bell System companies began to assume their shares of the system's debt, each had to be treated evenly. Obviously, the same was true when the assets were divided up.

Considering what faced the divesters, it was nearly incredible that they took on the job with expressed confidence. That they completed it on schedule—with many customer dislocations, but without total disaster in any area—is just about beyond belief.

Soon after entry of the decree by Judge Greene, a committee of BOC presidents came up with an essential to divestiture, the organization of the BOCs into new regional holding companies. The committee found it possible to split the existing companies into seven groups that were nearly equal in asset size and business prospects without carving any of the BOCs into pieces. Leaving all of the BOCs intact in their own territories saved incaluable time and expense such as avoiding changes in corporate charters and not having to revise regulatory or municipal permits.

A fact of human nature quickly rose to the surface. In the pre-divestiture Bell System, everyone worked for the same organization. Similarly, they were under the control of the same employer. Many top officials moved from the presidency of one BOC to another, all the while hoping to wind up in the AT&T "cabinet." Suddenly, in the minds of those who would manage the newly independent firms, AT&T was the "other guy." Divestiture would not happen for more than a year, but the manager's real responsibilities were to his BOC, and one of the seven as yet unnamed regional companies whose stockholders would own it.

Rapid-fire developments moved on toward the filing of a highly detailed plan of reorganization with the court. The 471-page document was accompanied by an equal number of pages of appendices. Submission of the plan of reorganization triggered another sixty-day period for public comments under the Tunney Act procedures.

In the meantime, the plan for the seven regional companies was outlined. Each would have approximately 10 percent of the assets of the combined Bell System, or as close to that number as was practical without splitting up the existing BOC boundaries. Chief executive officers for each were appointed. Mainly, they were the incumbent presidents of the largest BOC in each region, although in two instances, Southwestern and Pacific, the new region would consist of only one large company. One CEO, Thomas E. Bolger in the Mid-Atlantic area, was an AT&T executive vice president.

The CEOs lost little time expressing their independence, internally and a few times publicly. Judge Greene asked them for affidavits giving their views on the business viability of the new organizations they would head, and several of them raised problems they saw in the offing.

Chairman-designate Donald E. Guinn of the Pacific company, which would also include a Pacific Telephone subsidiary, Nevada Bell, disagreed strongly with the planned financial structure for his firm. Under allocation of the outstanding debt securities of the BOCs, all of the other regionals would have debt ratios of about 45 percent. Because of traditionally tight rate regulation in California, Pacific Telephone had done more of its financing through issuance of debt, and its debt ratio would be around 50 percent, or 54.4 percent if some outstanding preferred stock was considered as debt.

Pacific had another potentially serious financial problem. California's tough state commission had directed "flow-through" to present subscribers of reserves from deferred taxes, mainly the then-existing investment tax credit. Although majority opinion was that this was contrary to the intent of

Congress—the deferred taxes were felt to produce capital infusions at no cost to subscribers—lengthy and complex litigation had left Pacific Telephone facing a threatened back tax liability of some $2 billion.

All hands went to work on both problems. In January 1983, soon after Congress convened, legislation was passed relieving Pacific of most of the prospective tax liability. By April, under pressure from Justice and facing a potentially adverse order from Judge Greene, AT&T agreed to assume five outstanding debt issues of Pacific, bringing the BOC's debt ratio down to about 47 percent and lowering the price that future issues of debt capital would cost it. In complex but rapid transactions, AT&T also bought up the minority stock interests in several of the BOCs, including Pacific.

From the time of the filing of the plan of reorganization in December 1982, it had been evident that the use of the "Bell" name and the familiar bell-shaped logo was in dispute between Justice and AT&T, which wished to retain those valuable rights. Two of the regional CEOs expressed considerable concern on this score in their affidavits filed with Judge Greene. For a time, it appeared that the controversy might delay the tight divestiture timetable.

In March, AT&T, still unwilling to surrender the name and logo trademark, submitted an amendment to the plan of reorganization at the demand of Justice. It provided that if Judge Greene ruled in favor of Justice and the BOCs on the issue, the BOCs would have exclusive right to the trademarks in connection with their permitted local exchange and directory services. AT&T would retain the rights outside the United States. The BOCs, it was proposed, could use the name Bell if it were modified by a geographic designation.

Time was getting short for selection of names for the regions, so that the big job of changing the markings on everything, developing advertising campaigns, preparing for the issuance of stock, and numerous other details could get under way. Three of the regionals waited until they had reasonable assurance they could use the Bell name. The Mid-Atlantic group, from Pennsylvania and New Jersey south to Virginia, became Bell Atlantic Corp. The two companies serving nine states in the Southeast—at one time, they had all been under the Southern Bell banner—were now BellSouth Corp. Southwestern Bell simply retained its established name, adding a "Corp." to distinguish the holding company from Southwestern Bell Telephone.

The others decided to break with the past and adopt more futuristic names. The New York and New England companies combined their initials with "X" for future growth and became NYNEX Corp. The five single-state BOCs surrounding the Great Lakes, from Ohio to Michigan and Illinois, named themselves American Information Technologies Corp., familiarly Ameritech. The vast region from Nebraska and Minnesota through the

Rockies to the Pacific Northwest proudly heralded its future as U S West, Inc. The California and Nevada companies adopted a synonym for "intelligently managed progress" as the Pacific Telesis Group.

Changing the face of a basic U.S. industry which entered into every community, however small, continued as a frantic, round-the-clock operation. Gaining little public attention, but crucial to the entire process, was the time-consuming and detailed job of laying out every inch of territory in which the BOCs were permitted to provide service under the modified final judgment.

Each LATA (local access and transport area) had to be defined, usually consisting of a metropolitan area and the BOC's franchised territory surrounding it. Independents were associated with the LATA they adjoined. Because of the increasingly intense competition for intrastate long distance traffic, the formation of the LATAs was often a matter of heated dispute. The BOCs wanted LATAs as large as possible so they could furnish the connecting links between exchanges, while their competitors wanted small, numerous LATAs leading to lots of intrastate interLATA traffic.

At the beginning of March 1983, in the same week that AT&T filed another essential three-foot-high stack of documents, this time with the FCC, the entire timetable almost became completely unglued. It was announced that three Justices of the U.S. Supreme Court—only one more was needed to throw the whole process into Supreme Court review—had voted to take the matter under consideration.

Officials of thirteen states had challenged Judge Greene's decision to accept the modified final judgment, arguing that the authority of the states had been unlawfully preempted. The three Justices questioned the adequacy of the guidance given the court in the 1974 Tunney Act to accept the decree in the public interest. Justice William Rehnquist, joined by then-Chief Justice Warren E. Burger and Justice Byron R. White, said he was troubled by "the notion that a district court, by entering what is in essence a private agreement between parties to a lawsuit, [may] invoke the Supremacy Clause powers of the federal government to preempt state regulatory laws. The district court may well be correct, but I am not prepared to create a precedent in this court by summarily affirming its decision."

Nonetheless, by the narrowly adequate 6 to 3 vote, Judge Greene's ruling was summarily affirmed without further comment or hearing. The strength of the Supreme Court side questioning state preemption may have been a harbinger of the decision several years later overturning the FCC's preemption of the states when it prescribed uniform depreciation rates—the first time in more than thirty years that the high court reviewed a nonbroadcast communications case.

In the many-sided process, the FCC was working on the vast array of changes and transfers of licensing authority occasioned by divestiture. In some instances, AT&T took over permits originally issued to the BOCs, and the reverse was true in many others, depending on the plan of reorganization. The company answered one question which had been raised during pre-divestiture when it asked the Commission for the series of authorizations.

AT&T reported, as part of the severance of the formerly integrated system, that it intended to offer message toll service "on an end-to-end basis after divestiture. This means that customers who choose AT&T as their in-terLATA carrier would receive an end-to-end carrier charge for an MTS call, notwithstanding the fact that separate, unaffiliated entities would ac-tually be involved in providing the service. AT&T intends to assume the responsibility for arranging access with the local exchange carriers. AT&T's rates would be set to recover its costs, including access charge." In a market in which AT&T was the bellwether, and the others were competing against it, so would the competitors' rates.

In early July, Judge Greene issued a detailed, almost exhaustive 159-page order approving the plan of reorganization. He made some changes intended to "assist in moderating the pressure for local rate increases, whatever their source." Local rates have been a continuing source of concern to the judge, who obviously did not wish it said that applying the antitrust laws to permit a competitive society has cost the ordinary citizen more money in higher telephone rates.

These protective actions, the court said,

include the assignment of the "Bell" name and logo to the operating companies; im-position of the requirements that the operating companies be granted licenses to all patents required by the Bell System, along with appropriate sublicensing rights; the grant of authority to the operating companies to perform their own official [internal] services functions rather than to have to lease the necessary facilities from AT&T or other interexchange carriers; approval of the establishment of a central staff organ-ization which will give effective logistical support to the operating companies; im-position of the requirement that AT&T stand as the ultimate guarantor of the costs of equal access and network reconfiguration; and the approval of local access and transport areas (LATAs) which avoid substantial, costly network rearrangements.

AT&T was permitted to use the Bell name for the Bell Laboratories and in foreign operations, as had been proposed earlier.

The reality of divestiture was never more evident than in the subsequent brouhaha over a very basic monetary question, the guarantee of the equal access and network reconfiguration costs required by the whole process. Judge Greene directed that if after ten years or by January 1, 1994, the BOCs have not recovered those costs plus financing expenses, AT&T is to reimburse them the amount of any remaining deficit. A preliminary accoun-ting of the costs was to be provided by January 1, 1989.

AT&T raced back to the courthouse with fundamental questions about the equal access and network reconfiguration expense guarantee: What would happen if the BOCs did not try to recover all the costs, knowing AT&T would have to pick up the tab? Or what would happen if a state commission, delighted with the prospect of shifting its ratepayers' costs to AT&T, did not let the appropriate tariffs go into effect? Judge Greene quickly issued a supplemental ruling to the effect that AT&T would have to guarantee only those costs not attributable to the BOCs' failure to file appropriate access tariffs to recover the expenses, or a regulatory commission's refusal to let them become effective. Just what constituted proper equal access and network reconfiguration expenses led to another nose-to-nose confrontation, as described in the next chapter.

If the former Bell System partners were well along the way to being business rivals, the integrated service arrangements soon to be severed still left them some bonds. In mid-September, the Ameritech operating companies asked Judge Greene for a clarification of an operating matter not entirely clear from what had gone before.

The equal access process, of course, calls for each subscriber to select and designate a preferred long distance service provider among the competitors who operate at the customer's location. Ameritech asked for assurance that its belief was correct that the long distance calls of customers who did not express a preference could be routed via AT&T. It noted that undesignated calls would be a transitional phenomenon over the next several years—the FCC later made sure the transition period was shorter than that—but expressed its operating desire for the time being.

Judge Greene, in a ruling he later modified and which was subsequently superseded by the FCC when it took control of the carrier presubscription process, once again was a strict constructionist of the decree. While conceding that what was at stake was "billions of dollars annually," he ruled that the BOCs clearly were authorized under the decree to route undesignated traffic to AT&T if they wished. He took a dim view of such alternatives as the local exchange carriers' blocking long distance calls if no carrier had been designated, or of their insisting that the subscriber select a long distance carrier before his call would be handled.

All the pre-divestiture pieces somehow fell into place. Probably because the chaos which would have resulted was worse than any alternative, all of the participants made whatever accommodations were needed to ensure that the deadline, known to everyone as "1/1/84," would be met.

Even though divestiture was still a short time off, 1983 came to a close with a classic example of what things would be like between the former Bell system partners once their formerly close relationship had been severed. The last-minute flare-up took place during the 1983 Christmas season, when AT&T charged into Judge Greene's court with a request that he require Bell Atlantic and its subsidiary Chesapeake & Potomac Telephone, serving among other locations the national capital in Washington, to comply with the modified final judgment.

Bell Atlantic had taken a look at the provisions which said that AT&T owned the embedded customer premises equipment as of the date of divestiture. The federal government wanted to buy more than 200,000 telephones in government buildings in the Washington area, rather than pay monthly lease charges to AT&T Information Systems (AT&T-IS). Bell Atlantic's solution was fairly simple. It contracted with the government to sell the telephones to it just before the date of divestiture, while entering into a one-year contract for maintenance of the equipment, under which it would keep several hundred Bell maintenance employees as its own, rather than as employees of AT&T-IS.

Judge Greene determinedly took the position of a lawyer construing a contract. On the eve of divestiture, he issued a decision transferring to AT&T the contract with the government, along with the requisite personnel and the maintenance agreement. The Bell Atlantic companies, he reminded anyone interested, were wholly owned subsidiaries of AT&T and had no separate right of existence until they sprang from birth as wholly new entities on New Year's Day, 1984.

20

"CZAR HAROLD"

In the cold morning light of New Year's Day of 1984, FCC Chairman Mark S. Fowler and MCI Chairman William G. McGowan later reported the same reaction. They woke up, picked up their telephones, and discovered they still had dial tone. The telecommunications structure was different, but telephone service was still operating even with divestiture.

For most telephone users, some significant changes had occurred in their long-standing association with the local end-to-end service telephone company. The service itself, post-divestiture polls have indicated, has changed relatively little in quality. Who supplied the service and how the customer went about getting it had become a whole new process.

First, the customer had to learn that however much at odds with reason it seemed, in one sense his telephone company no longer had anything to do with his telephone. Perhaps he already had bought a new telephone instrument with all sorts of features in an electronics store or from his telephone company or AT&T Information Systems. Perhaps he still had the familiar embedded instrument in his house, apartment, or office; if so, he was leasing it from AT&T-IS or had already bought it from that entity. Either way, the instrument was no longer either the property or the responsibility of the local telephone company.

The customer had to be prepared to cope with a new set of problems and decisions when he found himself without the familiar dial tone. No longer could he expect a prompt visit by the friendly installer-repairman and the restoration of service at the earliest possible time. Now he had to undergo (perhaps while surviving wintry blasts at an outdoor coin booth because his home telephone was out of order) a quiz game. He might be convinced that the problem was in the telephone company's plant and not in his instrument; he had to be sure enough to be willing to pay for a nonproductive service call if the problem turned out to be in his telephone, and thus not the responsibility of the telephone company.

Just how customers survived without a fairly detailed knowledge of the ins and outs of divestiture, the organization of the telephone business, and the changed relationship of AT&T and the local operating company is nearly beyond comprehension.

There are two groups of businesses for whom the benefits of terminal detariffing and deregulation are obvious. Any organization using sophisticated high-tech data communications or Information Age systems is a net gainer. They can thank the competition era (which arrived before divestiture) for much greater innovation, broader availability of equipment, and better terms and prices from competitive suppliers. Among those users, the customer is always right—at least, much of the time. He does need to deal with increasingly complex technology without the aid of a sole source supplier.

Even bigger winners were the companies selling, distributing, manufacturing, and maintaining customer premises equipment (CPE), from PBXs to answering devices. Before deregulation they had to fight their way into the business and usually were met with negative attitudes of the telephone companies ranging from lack of cooperation to outright hostility. Now, they had open entry, and it was up to them to make it in a competitive world, although a lot of them could not.

For the great mass of telephone subscribers—the residential users and the small businesses looking mostly for plain old telephone service—the benefits of CPE deregulation turned out to be largely theoretical. To the dismay of many who rushed into the business, most customers regarded their telephone instruments as the most necessary and dependable appliance in the household or business, giving them instant access to the outside world. They were likely to keep the old ones as long as they worked, rather than change them with color schemes and interior decoration.

There was one economic benefit to the small users, existing for a very good reason. Buying the instruments saved some money over time against the monthly lease charges, because there was no longer any maintenance. Now that the policymakers have excluded all CPE as part of telephone service, the telephone company will not visit subscriber premises on call to repair or replace malfunctioning CPE. Instead, for example, the customer who buys a telephone from AT&T-IS receives a warranty from the company which advises: "For warranty service, return your telephone equipment within 90 days of the date of purchase to a location designated by AT&T. For information on warranty service, call the AT&T business office. If you send your telephone equipment, you must pay for all mailing costs. When you return your telephone equipment for warranty service, you must show proof of the date of purchase."

Users of long distance service have made out considerably better. This effect can be justified on economic grounds, in terms of shifting some of the expense from long distance to local users on a cost-causitive basis, but the change from old-style separations has had a substantial political impact. While the argument may be oversimplified, it is easy for a politician with a local or state constituency to argue that the FCC, largely in control of the access program, has tipped the balance in favor of interstate service. The FCC, in turn, contends that it is seeking to redress the economic imbalance which existed against interstate users, and that now the customers are paying the freight more in accordance with their responsibility for it.

In any event, as a result of the FCC's reallocation of costs and AT&T's own internal austerity measures, AT&T interstate rates had, by the beginning of 1987, plummeted by a cumulative 30 percent since divestiture. Since AT&T still had a major if not dominant share of the market, its rate reductions constrained its competitors to follow suit in order to remain competitive. The competitive atmosphere also was encouraging AT&T to offer optional service plans—from "Reach Out America" in several versions for residential long distance users to Megacom and SDNS (software defined network service) for big, sophisticated business customers. Countering, the competitors were going to customers with their own alternative or special service offerings.

Based on a somewhat rough assessment of what was being delivered at what cost, divestiture seemed three years after the fact to be a "wash" for the consumer who made a modest number of long distance calls each month. He would be paying more for local service, but with modest use of long distance service, the price changes offset each other. But for at least 50 percent of telephone subscribers who make two or fewer toll calls a month, there has not been the same consoling offset.

A last and more comprehensive look at where we are and where we are going will be reserved for the final two chapters of this book. It should be observed now that in the first three years after divestiture on January 1, 1984, the most sweeping changes were in the efforts of the former Bell System partners to maintain or enlarge their turf, and in the identity of the adjudicator of disputes which once had been settled by the chairman of AT&T.

Judge Greene would be happy, it can be assumed, to deal with the relatively simple internal disputes of the old Bell System. He cut his teeth on the heated AT&T/C&P contract fight over a government contract recounted at the close of chapter 19. That battle was mild compared with some coming up in the early post-divestiture years, and it may be predicted with confidence that the biggest fights were still ahead, to come in the late part of the 1980s and beyond.

For the fixing of policies which guide the management of the nation's telephone business, including what customers are to be charged for what service, whether access tariffs may go into effect, and how rules and regulations are interpreted, the interstate segment of the business looks to the FCC. Telephone companies are regulated, deregulated, and unregulated by decision of the Commission.

Beyond those critical operations of the established business, involving as they do many billions of dollars in ratepayer charges every year, are the aspirations of telecommunications managers and their stockholders. What lines of business, and on what terms, can these vast pools of capital—the seven Bell regional holding companies—enter or reenter? Unless reversed by the appellate courts (and this has happened just once at this writing) the answer lies in the hands of one middle-aged, mild-mannered World War II refugee. He is, of course, Judge Greene.

The heart and soul of the entire modified final judgment (MFJ), the control of the bottleneck monopolies found to be anticompetitive when they were owned by AT&T, lies in one section of the MFJ. Other than those businesses specifically permitted—exchange and exchange access telephone service, publishing and distributing yellow pages, selling new CPE—the Bell operating companies (BOCs) must gain a waiver from the court of jurisdiction before they can go into any new line of business. If their monopoly of local exchange service will or may impede competition in the new field, the answer is "no." In our judicial system, the answer is given by Judge Greene.

This vast power over the business activities and entries of the divested companies made Judge Greene, however carefully and benevolently he exercised the authority, the business dictator almost of all telecommunications pursuits. His penchant for administering strictly the terms of the modified final judgment he had entered under the laws of the United States as he construed them made him, in the eyes of those who wished a looser rein, "Czar Harold."

Unlike most absolute rulers, however, Judge Greene's powers are subject to higher court review and congressional alteration. A number of efforts at reversal have been made in both forums, almost exclusively by those aggrieved by his decisions on the scores of requests by the now-independent Bell operating companies to enter lines of business not specifically permitted by the MFJ.

Almost from the day of divestiture, "Czar Harold" has been called on repeatedly to interpret and relax the restrictions in the documents he approved. Many of the disputes involved, immediately or potentially, hundreds of millions of dollars in telecommunications business. Most were intimately related to the dramatic change in the business relationships of the companies which formerly constituted one nationwide integrated system. From the outset of 1984, AT&T and the BOCs have been consistently at opposite sides of nearly every business controversy which has arisen.

First of the monumental battles surfaced within two weeks after divestiture. Justice, well schooled for the upcoming battle by AT&T's inter-

exchange service competitors and representatives of the soon-to-be divested
BOCs, was ready in early January 1984 with a motion filed in the decree
court. It asked Judge Greene to require AT&T to permit the use by the
BOCs of its CCIS (common channel interoffice signaling) software system
for access and routing of "800" (inward wide area telephone service) traffic
to be handled by the long distance competitors.

Behind the arcane language of the motion was a fight over a business which
had amounted to $2.417 billion as early as 1982, and which has mounted to
upwards of $5 billion a year. Its use is familiar to millions of members of the
public and watchers of television commercials. Prospective customers of services being offered are urged to call toll free a number beginning with "1-800"
or "800" and often ending with a business slogan, such as "1-800-GET-
CASH." The long distance competitors were nearly frantic to compete for
this huge business, effectively belonging to AT&T alone when it retained all
rights to the use of the CCIS system.

The problem for the competitors, and BOCs wishing to provide equal access
for "800" service to them, was that "800" numbers dialed by subscribers
are different from the actual telephone numbers, with their built-in routing
instructions, at which the calls terminate.

As Justice explained in its motion, "In a competitive environment, the
BOC at the originating end of the call must be able to determine which of
potentially several carriers has been designated by the 800 subscriber—the
called party—to carry the call. Further, because the dialed 800 number con-
tains no routing information, the dialed 800 number must be translated into
a standard 10-digit POTS [plain old telephone service] number before it can
be switched by the terminating BOC to the appropriate access line."

AT&T first responded that the "NXX" numbering plan, the way "800"
service was provided before the development of CCIS, was available to the
BOCs if they wished to use it. The BOCs replied that they would still have to
engage in expensive development of their own database systems to provide
real equal access, meaning that they would have to scrap NXX for "800"
service use in two or three years when the database operation was ready.
Meanwhile, the long distance competitors protested: Their customers would
have to change the familiar "800" numbers which some had spent large
amounts of money to advertise.

AT&T's basic position was that "nothing in the decree or plan of reorganization was intended to give other carriers a free right to AT&T's
technology. The decree assures them equal exchange access, and thereby an
equal opportunity to compete in the marketplace. It does not entitle them to
avoid investment by taking for their own use AT&T's technology and interexchange facilities."

All seven of the BOC regional companies joined in a filing supporting the
Justice motion to compel AT&T to permit them to use CCIS. They said, "In

the absence of the 800 database system, they do not believe they can provide meaningful equal access to all interexchange carriers because AT&T will retain sole control of the facilities needed to perform the essential equal access functions, including translation." They observed that they were part of the Bell System when CCIS was developed, and thus helped pay for it.

The decision was not an easy one for Judge Greene, it appears, since it took him about a year to issue it in early 1985. He turned down Justice and the BOCs, reminding them that the plan of reorganization contemplated that the BOCs develop their own database systems. He said the ruling was "consistent with the decree's goal of establishing the [BOCs] as entities that are truly independent of AT&T."

Judge Greene, continuing, laid bare some of the philosophy he had followed in presiding over the dissolution of the world's largest business enterprise. He said the purpose of equal access "is a limited one: to remove barriers to entry and create a truly competitive environment for interexchange service. It is not the artificial creation of competition by enabling the interexchange carriers to share AT&T's capabilities and facilities to which they are not otherwise entitled. If the interexchange carriers wish to fulfill the decree's promise of flourishing competition, it will be up to them to make the necessary financial investments and to develop the appropriate technology."

He pointed to availability of the NXX system, noting that AT&T was offering the use of originating screening offices for that purpose. He added that the BOCs could obtain limited temporary waivers of the decree's ban on their providing interexchange service, necessary for them to use NXX and the AT&T originating screening offices for "800" services, for the asking. In 1986, most of them took him and AT&T up on these offers and started providing "800" service equal access to the competing long distance companies.

In less than two months after divestiture, the BOCs had accumulated a good pile of pending line-of-business waiver requests in the court. Justice came in with general comments in late February 1984, taking a generally skeptical view of allowing the companies to enter new businesses. In a position which changed considerably in three years—as will be seen in the final parts of this book—they asked Judge Greene not to allow the BOCs to divert their efforts from the main goals of providing exchange service and equal local access.

By late July, Judge Greene was ready with an order which generally bought the Justice arguments. The BOCs should concentrate on their primary job, he declared, while announcing that their entry into unrelated businesses would be "at a measured pace," "on a carefully controlled basis," and not "on a significant scale."

Just to make everything plain, he advised the BOCs that in addition to meeting the standards of the MFJ and the antitrust laws, their waiver applications must also comply with four added criteria. These are that revenues from all waiver activities will not exceed 10 percent of the regional holding company's total revenues; competitive activities must be operated by separate subsidiaries; the subsidiaries must do their own debt financing on their own credit; and Justice retains all "monitoring and visitorial" rights included in the decree.

A year later, his irritation with the continued stream of proposals to go into new businesses had not subsided. In the meantime, the BOCs had asked for several clarifications of the MFJ to permit them to go into additional operations, rather than filing straight-out waiver requests, on the basis that they were not sure whether their plans needed waivers or not. Judge Greene called for an oral argument on the subject in early August 1985.

The judge showed up for the session with a big stick in his hand. He previewed the order he would issue five months later with a stern opening lecture terming the large number of new business proposals the BOCs had filed an "incongruity" in the face of "many complaints" about local telephone service and rates. He viewed with little sympathy what he saw as an "almost frenzied desire" of the BOCs to get into new businesses.

In January 1986, Judge Greene denied the requests for clarification. He concluded that they were in effect motions for waivers and, if granted, would put the BOCs into the prohibited businesses of interexchange communications and information services.

At the same time, he reached the one conclusion which has been reversed on appeal. The BOCs had sought authority to provide cellular mobile radio services outside their normally franchised territories, but their proposals were styled in terms of extra-regional or extra-territorial exchange services. The court held they were restricted to providing all exchange services within their own exchange areas.

Judge Greene's decision was taken to the U.S. Court of Appeals for the District of Columbia by three Bell regionals on two basic grounds—the extra-territorial ruling, and U S West's contention that since the BOCs were parties to the consent decree, and were represented by AT&T as their parent, they are not bound by the document. The appellate tribunal rejected the latter contention quickly, and by early 1987, U S West was unsuccessfully seeking U.S. Supreme Court review on that point.

The Court of Appeals, however, did not go along with Judge Greene's finding that the BOCs had to stay in their own backyards. It declared that "neither the express language of the consent decree nor the circumstances of its formation will support the territorial restrictions imposed on exchange services by the district court. . . . There is no explicit or implicit geographic restriction of RHC [regional holding company] and BOC operations contained within the four corners of the consent decree. When we look beyond the language of the decree, we remain unpersuaded that the parties to the decree intended to prohibit provision of extraregional exchange services.''

There were other controversies involving the divested BOCs, usually with AT&T, and at least one multi-billion-dollar dispute was settled out of court. In early July 1984, Judge Greene was handed two issues potentially involving billions in assets and contingent liabilities. AT&T was demanding $659 million in "true up" payments from the BOCs, amounts it said were due after the January 1, 1984, accounts were reviewed. Not to be outdone, the BOCs came back with claims for about $3 billion on the ground that they were divested with "inadequate cash." Both claims were prospectively subjects for arbitration, but rulings by Judge Greene were requested.

Later in the same week, the BOCs asked Judge Greene to make AT&T comply with the plan of reorganization in connection with the former Bell partners' contingent liabilities, mostly antitrust case damage awards. The contingent liabilities were calculated on the proportionate basis of the assets going to each of the eight entities (AT&T and the seven regionals) upon divestiture. AT&T, the BOCs indignantly proclaimed, wanted to have its share calculated after its assets were written down by $8.857 billion at the close of 1983, to cleanse its balance sheet of overly valued items now worth much less in the new structure.

Fairly quickly, most of the BOCs and AT&T came to agreements on those disputes. They could not get together, however, on still another controversy which may involve considerably more than the $4 billion-plus hassles they had just settled.

As noted in chapter 19, AT&T had been tagged as the "ultimate guarantor" of the equal access and network reconfiguration costs of the BOCs over a ten-year period of divestiture, provided reasonable efforts were made by the BOCS and their regulators to defray them. This immediately led to a major struggle over the definition of those expenses which AT&T was to pay as of January 1, 1994, if they were not covered earlier.

Judge Greene split the outcome down the middle. He agreed with AT&T that the costs of advancing originally scheduled construction dates should not be included within the guarantee. But he went with the BOCs in their contention that administrative expenses, costs of construction included in pre-divestiture plans but actually performed later, and equal access costs incurred after the 1989 preliminary accounting deadline but before January 1, 1994, when the guarantee ends, should be covered.

In dealing with two very large and potentially troublesome transactions, Judge Greene demonstrated that he could accept something novel if the strict prohibitions of the MFJ were complied with. He was fairly relaxed in dealing with an agreement under which the Pacific Telesis Group acquired, in a $432 million deal, Communications Industries, Inc. CII was a conglomerate which operated in the cellular and mobile radio fields and manufactured equipment. As long as Pacific committed itself to dispose of such prohibited functions as manufacturing within a specified time period, Judge Greene went along with the proposal.

He was relatively easygoing, as well, in handling a transaction which seemed at first to be a real blockbuster. NYNEX Corp. entered into a complex arrangement giving it a conditional right—in effect an option—to buy 100 percent of the stock of Tel-Optik, Ltd., a projected new transatlantic carrier, within two years if it obtains the necessary waiver from the court. International communications originating or terminating in the United States fall under the MFJ prohibition against BOCs' providing interexchange service, and thus the BOCs need line-of-business waivers to be involved in providing such service.

Opponents of the transaction contended that it gives NYNEX an economic interest in Tel-Optik, and thus an incentive to discriminate in its favor. NYNEX, supported by the other BOCs, argued that the deal gave it no economic interest in Tel-Optik, but merely the right to acquire one later if it could meet the other conditions.

Judge Greene went along with the BOCs. "Not every expenditure made in pursuit of an acquisition target requires a waiver," he held. He gave Justice authority to approve such transactions in the future, on condition that the investment is relatively minor, there is no assurance that the transaction will be completed, and that the BOC lacks the ability or incentive to discriminate against the competitors of the target company.

Within little more than two years after divestiture, the BOCs and their parent regional companies, seeing other business opportunities available to firms of their size, resources, and assets, were tiring of trying to meet the strict antitrust standards laid down by Judge Greene. They began urging Congress to pass a law transferring administration of the MFJ to the FCC. In the process, they found ready allies in the leadership of the FCC, which had argued for some time that the rules governing crucial structural aspects of the industry should be administered on the Communication Act's public interest standard, rather than solely on the principles of the antitrust laws.

Senator Majority Leader Bob Dole introduced a bill to that effect, and Chairman John C. Danforth of the Senate Commerce Committee sought to push it along. Retracing their paths of a few years past, once again representatives of scores of closely interested organizations began intensive lobbying and public relations campaigns. This time, AT&T found itself with many companions in opposing the bill. The steam behind the measure seemed to have dissipated before Congress adjourned for the year, and its principal Senate backers lost their majority positions in the November elections when the Democrats regained control of the Senate.

The events of 1986 on Capitol Hill appeared to confirm an historic principle developed over the more than half-century existence of the Communications Act: It appears to be nearly impossible for Congress to make really substantive changes in the common carrier portion (title II) of the law. Presumably this is due to the complexity of the subject matter and the large number of diverse, substantial, and intensely involved interests with huge financial stakes. It is noticeable, however, that those conditions have applied as well in other difficult areas, and there still has been legislation. It seems that in communications there is always another way—usually through the regulators and the courts—to deal with the problems presented, so that Congress does not *have* to pass a law to address the issues presented.

In any event, proponents of easing the structural limitations on the BOCs began in the second half of 1986 to forget the legislative approach and concentrate on the possiblity of changing the decree itself. Their target was the triennial review of the operation of the decree, whether it was meeting its purposes, and whether changes were required, scheduled to be initiated by Justice and conducted by the court. They deluged the Justice Antitrust Division and Dr. Peter Huber, a Massachusetts Institute of Technology engineering professor serving as a consultant to write a detailed technical report on the state of the industry, with reams of data, information, and arguments. All were intended to demonstrate that the major local exchange carriers are in an increasingly competitive business, vulnerable to being bypassed by big users, including the interexchange carriers themselves, and in need of loosened bonds to permit a wider range of business activities.

Before the three-year review report was filed in January 1987, the BOCs received a hand from another and surprising quarter, although they reacted suspiciously to it as premature. AT&T proposed to Judge Greene that he modify his prior procedural orders for handling line-of-business waiver requests and let the FCC, rather than Justice, do the initial screening. Ostensible basis for all this was that the flood of waiver requests had overwhelmed Justice—more than 150 such requests had been filed by the close of 1986—and that premliminary screening by the FCC would allow Justice to resume its primary enforcement role.

Since AT&T was a leading opponent of the legislation which could have transferred MFJ jurisdiction from Judge Greene to the FCC, the proposal to give the Commission a more prominent role in dealing with line-of-business waiver requests gave some pause to observers. AT&T keynoted that under its proposal Judge Greene's jurisdiction would not be changed, and that he would continue to decide—after the predecision screening—whether waivers would be authorized or not. The FCC's role would be delineated by the doctrine of primary jurisdiction, looked at earlier in chapter 16.

Whatever he might have thought of the consistency of AT&T's views, Judge Greene took them seriously. When Justice filed its triennial review report, proposing draconian changes in MFJ restrictions on the BOCs, the court established a public comment schedule which asked for views on both proposals.

Changes in the decree proposed by Justice, recommended by a new generation of officials in the same Reagan administration five years after Bill Baxter and his colleagues had signed off on an argument establishing a firm regulated/unregulated dichotomy and a fence around bottleneck monopolies, were dramatic. Vertical integration would return to the BOCs with permission to enter manufacturing and marketing of telephone equipment. The competitive concerns of other information providers would lose in a new regime allowing the BOCs to provide such services.

In connection with probably the most significant of the restrictions on the divested Bell exchange companies—the bar on their furnishing interexchange service in competition with AT&T and the other long distance services they are required to provide non-discriminatory access to—Justice came down with a puzzling and perhaps impracticable recommendation. The BOCs, it was proposed, could furnish such service, but only outside their home regions. In-region interexchange service, Justice proposed, would be allowed only when the states removed barriers to local exchange competition and took the companies out of the bottleneck category.

True to form, Judge Greene took little time to dispose of Justice's recommendations. In a little more than five months after hearing oral argument (some 170 parties intervened, filing more than 600 briefs), Judge Greene handed down a 223-page decision in which he refused to relax any of the line of business restrictions imposed on the BOCs by the MFJ except as they applied to some non-communications-related activities.

He found that the proponents of change had not made a case to permit the BOCs to expand their operations into competitive interexchange long distance services, telephone equipment manufacturing, and the provision of information services.

In the judge's view, the BOCs were still the same monopoly bottlenecks that called for imposition of the restrictions as part of divestiture. As he saw it, the BOCs continued to possess the opportunity and incentive to use their monopoly powers to thwart competition in the relevant markets. In sharp disagreement with Justice, the BOCs and the FCC, he concluded that neither developments in technology nor the changes in market structure since divestiture destroyed the ability of the BOCs to "leverage their monopoly powers into the competitive markets from which they must now be barred."

21

OUTLOOK: STRASSBURG

In this joint endeavor, we have described those seminal events and forces that have transformed one of the most entrenched and essential institutions in the life of this nation—the telephone industry. We have attempted to give this accounting with objectivity and a minimum of second-guessing or revisionism.

Speaking for myself, it has not been easy to maintain a detached approach to much of the subject matter. This is because I was an involved participant in shaping some of the watershed policies. For much of my career in regulation, I lived with policies that were time-tested and of proved workability. Thus, I accepted with conviction the conventional wisdom that perpetuation of the monopoly solution—as epitomized by the Bell System— was the most effective way to satisfy the essential communications requirements in the United States.

But, in time, explosive technology in communications and allied electronic fields challenged this solution and cried out for institutional adjustments. Having become the principal steward of the FCC's policy-making apparatus, I was in a strategic position to effect the response. Thus, cautiously and incrementally, the FCC introduced the use of competitive market forces into selected dimensions of the regulatory equation. There certainly was no plan or expectation at that time by any of the policymakers involved that this modest augmentation of the traditional regulatory process would set in motion the kinds of forces which eventually engulfed and transformed the entire telecommunications infrastructure.

These comments may sound self-serving or have the ring of mea culpa. If they do, so be it. But they are intended simply to remind the reader of my background when offering the following personal observations and concerns about where we are and where we are going in the context of current policies and industry restructuring.

First, I am concerned that, notwithstanding the dynamism of the current technology, the unbridled application of laissez-faire economics cannot be expected to assure the most efficient employment of modern communications resources or the most beneficial distribution of their potential public benefits. Second, total dismemberment of the Bell System as fashioned by the modified final judgment may well prove antithetical to the inherent nature and economics of a nationwide, common user, universally interactive network. Let me elaborate.

With the impetus provided by unleashed competitive forces, technological innovation has been advancing at an accelerating rate, leaving its impact on consumer demand in all dimensions of local, national, and global communications. Personal computers and a wide diversity of other customer premises equipment have been proliferating. Computer terminals are going on line with private and public communications networks for interaction with in-house or external information services. Network structures are faced with the challenge of accommodating packet switching, satellite communications, fiber optics, microwave, cellular radio, and other technologies having a variety of new service potentials.

All the while, overshadowing the entire future of network operations are integrated digital transmission and switching systems with their promise of simultaneous carriage of the user's voice, data, and all other forms of intelligence.

Each of these technological advances has had, to a greater or lesser degree, commercial application in a competitive mode. Unquestionably, this will continue. But their real impact upon network architectures is yet to be felt, along with the full measure of their public and private benefits.

Realization of their potential benefits requires large infusions of effort and capital. In the case of private networks, only the most affluent and efficient business organizations or institutions have been able to cope with the perplexing factors of systems analysis, identifying and choosing among hardware and software systems and suppliers, and making the investment involved in implementing those decisions.

To some extent, the burden has been mitigated in the data transmission sectors by the value-added network providers such as Telenet, Tymnet, and others. They have built packet switching networks serving up a mix of modern data transmission capabilites to satisfy a diversity of consumer needs. But to upgrade the public telephone network by integrating these advances in technologies for general commercial application to voice, data, facsimile, video, and other modern-day use on a demand basis takes a lot more.

Huge expenditures of capital and human resources are of course required. More important, their effective integration on a nationwide scale into existing network architecture and its components takes coordinated planning and action involving the entire telecommunications infrastructure. This in turn requires formulation of national, if not global, standards to facilitate optimum design and performance of networks with technically compatible transmission, switching, and terminal components. It demands concerted interaction among local and long distance telecommunications service providers, the information community, and relevant developmental and manufacturing resources of the nation. Until now there have been no encouraging signs that either government or industry is disposed to launch the necessary initiatives to deal with the problem.

Absent such unified and coordinated efforts, there is likely to be a continuation of uncertainty and confusion in the definition and realization of common or complementing goals. This increases the risk of a diversity of autonomous, disparate systems, both public and private, lacking compatibility and connectivity—the touchstones of network efficiency. The situation would not appear conducive to promoting the national interest in modern, efficient telecommunications.

The dedication of today's policymakers to the reliance upon marketplace solutions, in tandem with the dismemberment of the Bell System, appears to me to be working against the grain of the unified and coordinated effort and the commitment of financial resources that are needed in today's environment. What we have are intensifying rivalries within and among all sectors of the industry, including the new "baby Bells" and their progenitor, AT&T. It has become a case of every man for himself and his immediate profit-and-loss statement.

But, to me, the greatest cause for concern is the fragmentation of responsibility for the basic switched network, not only in terms of its day-to-day management, but in planning and development of technical and operational advances.

Historically, these responsibilities were centered in AT&T. With all of its shortcomings, it did the job of network architect, manager, and coordinator with general public approval. But that institutional framework is behind us, and the fragmented infrastructure, particularly with its multiplicity of intercity carriers, must now regroup in some fashion to achieve the same results under much more demanding and complex conditions.

It is seemingly ironic or aberrational that the ultimate restructuring of this crucial industry could occur with virtually no input from the Congress or its policy-making surrogate and trustee of the public interest, the FCC.

Unfortunately, the architecture of the modified final judgment (MFJ) was dictated by an academic, William Baxter, who had the last word on behalf of the government. He was convinced that it was antithetical to antitrust principles and sound economics for AT&T to remain in common control of the exchange and interexchange operations of the Bell System.

In his view, interstate long distance service was no longer a monopoly, now that MCI, Sprint, ITT, and a host of others were offering competitive alternatives. It followed, therefore, that AT&T, whose market dominance was now being challenged, had to divest itself of these local bottlenecks upon whom all the competitors, including AT&T, depend for their origination and termination of long distance calls transiting their facilities. This became the MFJ's controlling premise.

However, it is doubtful that the bona fides of this new interexchange competition was ever closely examined by Baxter, AT&T's Brown, or the final arbiter, Judge Greene. For it would have been readily apparent that the validity of this fundamental premise for dismemberment was far from established. This is because what passed for competition rested on artifically contrived and disparate economics. Obviously, AT&T's interexchange competitors were able to penetrate the lucrative long distance telephone market with their alternative offerings only as long as they were able to obtain dial-up access to the exchange operations of the local telephone companies at a fraction of the costs incurred by AT&T for its exchange access.

As recounted in chapter 14, preferential treatment was deemed justified by the contrived line-side interconnection that the other long distance carriers had been able to obtain from the local companies, in contrast with the hard-wired trunk-side interconnection that was endemic to Bell System network design and operations.

But as contemplated by the negotiators and terms of the MFJ, it was to be only a matter of time before the ersatz long distance services, with their costing and pricing advantages, would depart from the scene when the competitors of AT&T were accorded the same interconnection arrangements enjoyed by AT&T. They would all be given the same trunk-side automated interfacing with the exchange networks, at equivalent charges.

Experience has demonstrated that as AT&T's competitors have achieved this parity of access with the higher costs that go with it, they have encountered increasingly tough going. Their numbers have declined. Their profit margins have diminished or disappeared. Price differentials which were the principal sales feature of the ersatz services have almost disappeared. Market shares of the competitors have ceased to expand.

At the same time, general rate levels for long distance calling have been dramatically reduced, with AT&T taking the lead. This is largely the result of actions taken by the FCC to assign more of the costs of local plant directly to all subscribers, whether or not they make long distance calls. With its predominant share of the long distance market, AT&T has been the principal beneficiary of this shift in local cost recovery. In addition, the new competitive climate has compelled AT&T management to strip the company of those costly layers of operating fat that grew from generations of complacency and indifference typical of the monopolist. Under pressure from AT&T's rate cutting, its competitors, when touting their public offerings, must now rely less on price differentiation, with greater emphasis on the quality of their services and flexibility rate structures and billing arrangements.

It may be argued, therefore, that the consumer is deriving some of the conventional benefits of a pluralistic marketplace. But, in my opinion, it is questionable whether multiple suppliers of interexchange facilities do not simply contribute to discordant network planning, overbuilding, and redundancy without any significant compensating social or economic gains.

As I interpret the signposts, they all point to a return to the concept that long distance service, like local telephone service, is inherently a natural monopoly; that duplication of interexchange networks is at odds with the economies of scale; and that multiple suppliers of interexchange facilities do not advance the public interest in a cohesive, fully interactive nationwide common user network.

In this regard, keep in mind that the crucial part of any long distance network in terms of cost and service performance is its local distribution component provided by the local telephone monopoly. Every long distance call must transit that exchange plant for the entire duration of the call.

In short, the local telephone company does not simply provide *access* to a long distance service; it is a long distance service provider in its own right. In large measure, those local facilities and their capabilities determine the quality of a long distance connection, as well as many of the features and conveniences that can be engineered into the network service. It is for these reasons that about 50 cents of every long distance revenue dollar goes to the compensation of the local telephone company as a joint participant in long distance service.

If I am correct in my reading of the signs, a reading shared by many other knowledgeable observers, the nation is moving toward a return to some form of regulated oligopoly, cartelization, or monopoly in interexchange carriage. If this is in fact the trend, it will have some positive benefits for the national interest. It will lessen uneconomic duplications and counterproductive fragmentation in the communications network. It will contribute to more unified, integrated planning and programming of a modern, versatile nationwide common user network.

To achieve a unifying approach and cohesiveness in today's environment of network fragmentation, laissez-faire, and intense rivalries may be expecting too much. I take some comfort, however, in the judicious refusal of Judge Greene, as noted in the preceding chapter, to modify the MFJ so as to permit the BOCs to expand their role in long distance telephone service and to enter other competitive lines of business. Thus, at least for the short term, the chaos and stresses already besetting the current telecommunications infrastructure will not be exacerbated by the BOCs' use of their existing monopoly powers and earnings to distort emerging markets.

The stakes are high for all players. But it is the consumer at large who stands to lose if industry and government do not have the collective will and wisdom to meet the new challenges.

22

EPILOGUE: HENCK

As we have seen, divestiture and its close associates, competition and deregulation, have not been unalloyed disasters. They have been, however, disappointments to several of their major constituencies.

It has often been pointed out that divestiture amounted to a private agreement between AT&T and Justice, put into effect with some changes by a single federal judge. Since then, AT&T has been unable to fulfill its commercial expectations. At the same time, the deregulation it anticipated has been very slow in coming. AT&T seems to have paid a large price for very little.

Now Justice, in the short space of five years, has proposed virtually to abandon the basic premises on which the consent decree was based. Although the metamorphosis of the local exchange carriers from monopoly bottlenecks to competitors for local exchange service and access remains almost totally prospective, Justice has recommended that the Bell companies, if they meet nondiscrimination conditions, be permitted to move far closer to the vertical integration stance of the former Bell System.

The main prohibition then remaining on the Bell operating companies (BOCs) against providing interexchange or interLATA toll service, would be temporarily preserved more for practical than for equitable or legal reasons. At first, Justice recommended that the BOCs be permitted to provide interexchange service outside their own franchised territories, unless and until local regulators allowed free entry into exchange service. Later, the department withdrew that suggestion, citing the administrative complexity already mentioned by other commenters.

An infallible crystal ball would make most pursuits in life, including writing epilogues, far more convenient. No one at this writing can foretell what action Judge Greene will take in reviewing the decree's effectiveness in meeting its goals, and the proposals for change. Even less confidence could be put now on a forecast of what appellate courts might rule in the almost certain litigation to follow.

But the provisions of the modified final judgment, however they may be revised, constitute only one of the forces which will shape the future relationship of the U.S. telecommunications industry to its customers. Business, economic, and political history, along with present trends, provide some substitute in analysis for the impossible dream of a reliable ouija board.

AT&T's economic future, and thus to large measure that of the interexchange telecommunications industry, is inextricably bound with the progress of deregulation. There should be brighter days ahead for the stockholders of a company whose bottom-line financial problems in the first several years of divestiture may turn out to be of a short-term nature.

Several years have been needed for AT&T to position itself for its role in competitive interexchange service. Its earnings have been battered by huge write-downs of obsolete plant and equipment it carried into the postwar period and which were demonstrated to be of little use in the new environment. The company was staffed for a monopoly-style operation, and there again it has taken several years to trim unnecessary parts of the structure.

Meanwhile, the implicit promise of deregulation has been slow in materializing. It takes little analysis to understand the company's problems in fighting almost totally unregulated competitors when its new service offerings must be scrutinized by regulators, and when more than half of its operating expenses—charges for local connections—have been controlled by a form of regulatory intervention which gave its competitors 55 percent discounts.

At the same time, AT&T's early post-divestiture difficulties may serve as an object lesson to its former partners, the BOCs, now thirsting to emulate their parent's prior example. It took AT&T nearly three years after divestiture to decide to concentrate on its core businesses and move away from certain largely unrelated fields, most notably computer hardware, into which it had plunged. There are many examples in other businesses of companies which have decided relatively recently to focus on the things they do and manage best and to withstand the lure of new ventures and the establishment of conglomerates.

Justice surprised many observers by its recommendations in early 1987 to reconstruct in seven entities much of what had been torn down on the national level. Its proposals were, of course, premised on the belief that competition is gaining in the local marketplace, so that the individual telephone companies may no longer in the future be considered bottlenecks.

It may seem a little strange that the legal relief to be offered from the strictures of the modified final judgment was proposed, not for AT&T which has suffered the most financially, but for the regional companies while they were consistently splitting their stocks and increasing their dividends.

The key to the future, however, will not rest so much in the financial health of the affected companies—though that is obviously of great significance —as in the reaction of the public and its representatives to what is being delivered to them, and at what price.

Little value can be found in second-guessing the decisions which have brought the nation's telecommunications structure to this point. It is much too late and serves no practical purpose to consider whether AT&T was treated fairly at many points during the events recounted in the first twenty chapters of this book. Again, the crucial issue is how most telephone customers have fared.

In that regard, it is important to consider whether the conditions which underlay the past decisions of policymakers will continue. In one important area, the national political arena, a student of practical political science may conclude that they will not.

During earlier times, the critical question was whether government regulation could be an adequate substitute for the consumer benefits brought about by competition. Now, it is whether competition has delivered, or will deliver, enough of those advantages to replace those lost in the wake of deregulation and the departure of one-stop, end-to-end monopoly service.

Even the most casual reader could not fail to notice how, in the preceding two decades, policymakers led by the FCC took virtually every step possible to protect competitors. The three-year head start given non-telephone company domestic satellite operators was typical. The Commission, in its first computer inquiries, even went outside the communications industry to protect data processing against telephone company competition.

Obviously, if one believes competition per se is in the public interest, there must be competitors. Companies are not born full grown, ready to be providers of a significant market share. To foster competition, policymakers had to foster fledgling competitors just leaving the nest.

Still, despite the findings of the Supreme Court in the three circuits case more than three decades ago, the tendency was to consider competition good for its own sake. This may be the most important element of the present scene. For the first time in telecommunications history, we now have the opportunity to learn and to gather empirical evidence on whether competition in all markets serves the interests of most consumers.

We have been brought to this point by a largely laissez-faire school of political philosophy which holds that marketplace forces, less fettered by government, constitute the best means of regulating business for consumer benefit. This line of thinking began to reach its peak during the Carter administration and has climbed to a zenith during the two Presidential terms of Ronald Reagan.

Those years have given economic theory its laboratory and test bed, to consider if in fact it is better than regulation for the large majority of consumers. Perhaps, in a decade or two, we will know. But actually, we may not really *know*. We will have strong beliefs—and they are likely to be controlling, politically—based on our orientations. We have already seen that the present era has produced both winners and losers, putting the big toll users and high-tech companies, as well as numerous politically potent telecommunications companies, on the winning side. Average telephone customers, and the public officials most reliant on their votes and support, have not done so well.

It is not my role, or within my capacity, to predict the outcomes of the national elections in the late 1980s and early 1990s. It is not difficult, nonetheless, to remember that national political trends are cyclical. Those whose mindset turns toward greater government intervention may well be in control, nationally and increasingly in the states, within a few years. The great deregulatory experiment may be reaching its close, in favor of a new form of regulation.

A belief in cyclical trends in economics and politics, and a long-held view that in time regulation will perform something of a comeback, does not lead one inexorably to turn back the clock to the early New Deal. After what has happened, it is hard to anticipate government control of all forms of communications service by a rate-of-return-regulated monopoly carrier. Instead, what has happened in the past twenty or twenty-five years seems to point toward a mixed bag, for political, social, economic, and technological reasons.

Consider what has happened in airline deregulation. It may turn out to be the classic case of too much, far too soon. Some passengers, dedicated consumers who can shop for and meet the qualifications for special discounts, or those frequent flyers in a position to take advantage of cumulative mileage plans to gain occasional free flights, may give deregulation fairly good marks. Many others, burned by suddenly withdrawn offers, difficulty in reaching smaller cities, overbooking, and constant delays caused by overscheduling in peak hours to popular destinations, are in the opposite camp. The gradual return to some regulation may start with a recent proposal that the government regulate deceptive promises in airline advertising.

In telecommunications, those best equipped to look out for themselves have benefited from deregulation and competition. If the big, sophisticated, high-technology users have special and advanced needs, and the ability to hire specialists to ensure that they are getting the most for their money, and thus have posted gains, there is no point in changing things for them. If the new structure allows them to remain on the network, and if filling their requirements makes a contribution to the overall system, it does not appear that anyone has suffered.

The beneficiaries, in fact, may include the small users whose vulnerability to bypass—the loss of the most productive contributors to the common carrier network, leaving the little fish to try to survive—has been the most potent argument favoring the wide-open marketplace.

Nonetheless, it has been noteworthy that as the industry structure accommodated itself to new methods of supply, with the big interexchange carriers offering redeployed services of advanced rate and system design, concern about the bypass threat has waned to some degree. This has come about while a variety of "band-aid" interim access pricing measures, proposed to the regulators, have been sent by them back to the drawing boards for further examination. The desperately sought flexibility in the sytem has come from outside the regulated access charge structure, not from within the elaborate schemes which regulators have devised hoping to achieve a reputation as deregulators.

What all this seems to mean is that regulation cannot be all things to all people. No elaborate government-developed program, or one initiated by a carrier and approved by a government agency, should be needed to protect the big, technologically oriented users. They have the economic and technical muscle to make their own arrangements, and they may be expected to wind up with suitable services meeting their needs at reasonable (and probably discounted) rates.

But very much on the other hand, residential and small business customers are essentially dependent on some government entity, national or state, if they are to avoid being shortchanged. The competitive market has done very little, if anything, for them, and it can hardly be expected to do much in the future without some change in direction. Consumer organizations have not demonstrated the analytical and political skills necessary to help. The voice of the common man as a telephone user may again be heard only if and when some perhaps modest reincarnation of the New Deal takes place in the 1990s.

We may well have, therefore, the political, social, and economic bases for a return to a stronger system of regulation. Customers are likely to question whether deregulation and competition in all of the formerly regulated lines of business—and divestiture in telecommunications—have given them anything in exchange for the suppliers' freedom to charge rates closer to what the market will bear.

Socially, basic utilities still bear the stamps which gave them the name of *public* utilities. If telephone rates are to some extent social instruments (and every seriously considered access charge plan has included provisions for lifeline service and assistance to high-cost rural or sparsely populated areas), economic theories are likely to be mixed with social concerns.

Increasingly, it appears to me, this will come about by applying the direct controls of government regulation to rates for basic individual local services and plain old telephone service in the interexchange field. Telephone companies would trade off willingness to undergo tight regulation of rates for the services most used by residential and small business customers for the freedom to make the financial and technological arrangements to attract and retain larger users.

To a major degree, we are discussing a task for state regulation. Some state governments, most notably in the western half of the nation, have moved recently toward various forms of deregulation. Where intrastate toll service is truly competitive, deregulation or detariffing may stick. To the extent that local exchange service charges for small users are removed from control, state officials and legislators are likely to hear about it, long and loud.

It becomes progressively harder to justify rate-of-return regulation of AT&T, alone among several strong competitors to be favored by such treatment. Although AT&T still had nearly 80 percent of interLATA message toll telephone volume at this writing, and about 60 percent of all toll traffic including intraLATA, its customers include a disproportionately high number of small users. At very small toll volumes, the expense of record keeping and billing alone wipes out any profit from the customer's calling. No one except a company stamped by policymakers as the "carrier of last resort" would seek such business, and AT&T's competitors have turned strongly toward the business market as their mainstay.

As long as AT&T is under a de facto mandate to continue nationwide rate averaging, and the market is intensively price competitive between major population centers, the practical necessity for continuing to put the dominant label on AT&T and regulate its earnings from basic telecommunications services is obscure. The FCC will retain that authority under the Communications Act, but the suggestions of some of its policy analysts for rate caps on core services seem to be a more viable way of exercising it.

There is another, and perhaps equally pressing, requirement to continue the federal presence in telecommunications regulation. This is the perceived need for maintenance of control of the vital price relationship between interLATA and interexchange calling and the local exchange connection. To the federal agency, the local exchange companies which once were part of the Bell System remain dominant as the gateways to customer access. The modified final judgment, however it may be revised, presumably will still require nondiscriminatory local access. But the level of those nondiscriminatory rates is probably too important an element in the finely tuned regulatory structure to be left to marketplace forces.

As a matter of a couple of relatively simple facts, the federal presence then is likely to go on. Barring the unlikely prospect of a major congressional rewrite of the Communications Act—an unlikelihood which seems to dim even further as the new scene's outlines become firmer and clearer—

that wondrously elastic piece of legislation remains intact. Within my working lifetime, under its provisions, at first common carriers were regulated fairly strictly. Then controls were lifted from most of them. Now there is a strong body of opinion that legal precedents would permit even further deregulation. Almost anything goes.

As a consequence, there appears to be nothing in a statutory sense to prevent some controls from being reapplied, and all of this can happen without substantive change in the FCC's governing law.

At some level, the subscriber line charges imposed by the FCC make sense to all but full-blown populists. In early 1987, the Commission appeared to have arrived at a reasonable compromise by capping the future level of those charges at $3.50 a month, to follow several progressive increases over the $2 level then in effect. As a further opportunity for congressional intervention, there were to be additional reviews of the situation before two annual raises of 60 and 30 cents a month, respectively.

The accuracy of speculation about the future, of course, depends not only on the market but how that market treats its customers, and how they react. Customer convenience could, unhappily, become a thing of the past. It may be that bagging your own groceries in supermarkets and dishing up your own food in restaurant buffets will become the accepted norm. Perhaps most travelers will accept relying on marketwise travel agents in lieu of direct dealings with the airlines.

But that kind of wide-open competition does not seem to be in the cards for telephone services. For the small user, it hardly appears reasonable or practical to expect telecommunications consultants to play the role of travel agents. In the market itself, interLATA telephone service is moving toward an oligopoly. At the local service level, competitive alternatives are probable only for very large customers. Even for them, especially as their businesses depend on communication with individual customers, dealers, and suppliers, continued reliance on backbone local systems is likely to be in the offing.

If social, market, and political factors are getting into place for a modest comeback of regulation some time before the end of the century, then so are the technological elements of the telecommunications industry.

Conferences of futurists have led to some consensus that fiber optics will revolutionize the entire industry. We are told that there will be so much capacity that telephone rates will be "postalized"—the same rate for a transcontinental call as for a local call. This will, it is said, join with other advances, including the integrated digital services network, to lead to a complete merger of the telephone and the computer. The same device will be used for both communications and data processing, it is forecast.

From this corner, a dissent. Technology alone does not drive the course of history. Instead, human, social, and economic factors combine to bring about the use, non-use, or enhanced use of technology which often has been present for a long time. There is no question that technology is the critical first step. But it is the other factors which control timing, extent of use, and sometimes whether much of any use at all is made.

A new generation weaned on personal computers may accelerate a partial merger of computers and communications. Nevertheless, it is hard to visualize the disappearance of the telephone used for voice communications (and, of course, linked to a system readily available for data transmission). If both separate and joint use can now be made to work cheaply and efficiently, and those factors will increase with the rise of the integrated services digital system and other developments, why destroy the individuality of the partners?

Any attempt at postalized rates would run into even higher barriers, whatever technology may promise. Assume for the sake of discussion that regulation or its lack would permit such a move. The futurists appear to believe that all customers would be overjoyed at the opportunity to make a transcontinental call for the price of a local conversation. They forget that the big question would then be: If we pay the same price for a local call as for one that spans the continent, then isn't the local rate too high, and are we not subsidizing the longer haul user? It must be remembered that extensive use of satellite tranmission, whose costs are essentially distance insensitive, has not revolutionized pricing.

Human beings simply do not always behave as futurists, technologists, and economists believe that they should.

Should things turn out the way suggested, with stronger regulation of general public services and a light or non-existent hand on the carrier offerings designed for the big users, the earlier regulators who envisioned some competitive entry into selected services to enhance the effectiveness of regulation may have been remarkably prescient.

At first blush, it may appear that their early policy-making moves brought about, slowly but surely, a total revamping of the industry structure—that relatively minor intercession made the patient a little bit pregnant. The analogy goes further with the subsequent efforts to protect "embryonic" competition. But it is well to recall that the only right to life enjoyed by the competitors was that of providing improved service to the public, or equivalent service at lower rates.

The critical question for the future is whether the nation's telephone customers will benefit, not just whether the competitors can survive.

The time has come to reflect on the past for just a moment. My contribution to this joint effort (the analysis of a close observer whose thoughts were mainly those of the many news sources who went beyond what their jobs demanded and cleared away some of the underbrush left behind by the onrush of developments) was that of mirroring people who understood what was happening and were willing to share that understanding.

In no particular order, a debt of gratitude must be expressed to the following, some of them no longer with us: John deButts, Charlie Brown, Bill Ellinghaus, Bill Mullane, Dave Byers, Charlie Jones, Jack Scanlon, Ernie North, Fielding Woods, Jim Chance, Elmer Pothen, John Fox, Ken Heberton, Dick Callaghan, Paul Henson, Mal Lothschuetz, Dan Fisk, Sam Shawhan, Gaylord Horton, Merv Alexander, Don Hirt, Bill Friedman, Gorman McMullen, Roy Baker, Asher Ende, Bob Jones, Joe Fogarty, Dean Burch, Paul Darling, Rosel Hyde, Larry Darby, Dick Wiley, Everett Krueger, Ben Wiggins, Paul Rodgers, and my co-author, Bernie Strassburg. Of those inadvertently excluded, we ask forgiveness.

Special thanks go to John Connarn, who urged me to continue this project at a time when it was nearly abandoned. Recognition should be given to my daughters, Kathryn Parker (who made a major suggestion about the title of this book) and Joanne Nierle, both AT&T employees, and my son, Bill, who worked for AT&T before he attended law school. In a world in which I communicated mostly with the telephone industry's "chiefs," they taught me to understand the viewpoint of its "Indians."

If it is ever to be published, a book must end somewhere. Writing in the bright sunny days of early summer 1987, when the landscape is as full of promise as the dynamic telecommunications industry, I am reluctant to end the evolving story of the past half-century. The temptation is strong to ask for a few days' extension of the deadline, to gain one more critical fact or, with the greatest of good fortune, some added insight.

But a deadline extension would no more bring the final answers than turning back the clock an hour would freeze the dawn of a new day. Telecommunications people must keep on looking to the challenge of the unknown future. Whatever it is, it will be engrossing, productive, and worth waiting for.

REFERENCES

The following listing refers (often by popular name) to legislative, administrative, and judicial proceedings addressed throughout this book. It does not purport to be a comprehensive documentation, but selectively identifies the more significant actions that are reflective of communications policymaking since 1934.

CHAPTER 1

Communications Act of 1934, as amended, 47 U.S. Code, *et seq. Special Telephone Investigation by FCC.*

Public Resolution No. 8, 74th Cong., 49 Stat. 43; Proposed (Walker) Report on the Investigation; Report of the FCC, 76th Cong., 1st Sess., H.Doc. No. 340

CHAPTER 2

Western Union/Postal Telegraph Merger, 10 FCC 148(1943)

Smith v. Illinois Bell Tel. Co., 282 U.S. 133(1930)

NARUC-FCC Report on Separations, April 28, 1947

CHAPTER 3

Cammen v. AT&T, 2 FCC 351(1936)

Telephone Recording Devices, 11 FCC 1033(1947)

Telephone Answering Devices, 18 FCC 644(1954)

CHAPTER 4

Hush-A-Phone Corp. v. AT&T, 20 FCC 391(1955); rev'd 238 F2d 266(1956); Decision on remand, 22 FCC 112(1957)

CHAPTER 5

Report on NARUC-FCC Toll Rate Subcommittee on Toll Rate Disparities, July 1951

"Jurisdictional Separations"—for documentation, see Proceedings of NARUC Annual Convention for 1950 (Phoenix Plan), for 1951 (Charleston Plan), for 1954 (Modified Phoenix Plan), for 1964 (Denver Plan)

AT&T and Western Union Private Line Rates, 34 FCC 217(1963); *aff'd Wilson & Co. v. U.S.*, 335 F2d 788(1964), remanded on other grounds, 382 US 454

CHAPTER 6

"1956 Consent Decree"—U.S. v. Western Electric; 1956 Trade Cases (CCH) ¶68, 246(1956)

"Celler Hearings"—Report of Antitrust Subcommittee of H.R. Judiciary Committee pursuant to H.Res. 27, 86th Cong. 1d Sess.

CHAPTER 7

"Comsat Legislation"—Communications Satellite Act of 1962, 47 U.S. Code 701, *et seq*; Senate Rep Nos. 1319, 1584, House Rep No. 87-1279, 87th Cong. 2d Sess; February 7, 1962, letter from President Kennedy to Congress transmitting proposed satellite bill

"Cables v. Satellites(TAT-4)," Memo Op. and Order, 37 FCC 1151(1964)

CHAPTER 8

"Above 890 MHz"—Allocation of Microwave Frequencies Above 890 Mc., 27 FCC 359(1959), recon. 29 FCC 2d 825(1960)

"Telpak Rates"—37 FCC 1111(1964), *aff'd sub nom. Amer. Trucking Ass'n v. FCC*, 377 F2d 12(1966), *cert denied* 386 U.S. 943(1967) (Telpak A & B)

"Hi-Lo Rates"—55 FCC 2d 224(1975), *amended on reconsideration* 58 FCC 2d 362(1976)

"FDC v. LRIC"—Memo Opin. and Order, 61 FCC 2d 587(1976), *recon.* 64 FCC 2nd 971(1977), 67 FCC 2d 441(1978); *reviewed sub nom, Aeronautical Radio, Inc. v. FCC.*, 642 F2d 122, (1980), *cert denied* 451 US 920 and 976(1981)

"Telpak Sharing"—23 FCC 2d 606(1970), remanded 449 F2d 453(1971); 31 FCC 2d 674, 32 FCC 2d 619

CHAPTER 9

"Carterfone"—Use of Carterfone Device in Message Toll Telephone Service, 13 FCC 2d 420(1968), *recon. denied* 14 FCC 2d (1968)

"MCI"—Microwave Communications, Inc, 18 FCC 2d 953(1969) *recon. denied*, 21 FCC 2d 190(1970)

CHAPTER 10

"Investigation of Charges of ATT Charges for Interstate and Foreign Communication Service"—9 FCC 2d 32(1967)

"Second ATT Rate Investigation"—38 FCC 2d 213(1972), 41 FCC 2d 389(1971)

"Western Electric"—54 FCC 2d 1(1977)

"Ozark Plan of Separations"—See Proceedings of NARUC 1970 Annual Convention; 26 FCC 248(1970)

CHAPTER 11

"Post-Carterfone Tariffs"—Memo. Ops. and Orders, 15 FCC 2d 605(1968); 18 FCC 2d 871(1969)

National Academy of Sciences Computer Science and Engineering Board Report of a Technical Analysis of Common Carrier User Interconnection (1970)

"CPE Certification/Registration"—Memo Op. and Order, 35 FCC 2d 221(1975); *First Rept. and Order*, 56 FCC 2d 593(1975); *Second Rept. and Order* 58 FCC 2d *Third Rept. and Order* 67 FCC 2d 1255(1978)

"Telerent"—Memo Op. and Order, 45 FCC 2d 204(1974), *aff'd North Carolina Util. Comm. v. FCC* 537 F2d 787(1975)

Mebane Home Tel Co., 53 FCC 2d 473(1975)

"Computer I"—Interdependence of Computer and Communications Services, Notice of Inquiry, 7 FCC 2d 11(1966); *Tentative Decision*, 28 FCC 2d 291(1970); *Final Decision*, 28 FCC 2d 267(1971); *GTE Service Corp. v. FCC*, 474 F 2d 724(1973); *Dec'n on Remand*, 40 FCC 2d 293(1973)

"Computer II"—Second Computer Inquiry and Proposed Rulemaking, 61 FCC 2d 103(1976); 61 FCC 2d 103(1976); *Supp. Notice*, 64 FCC 2d 771(1977); *Tentative Decision and Further Notice*, 72 FCC 2d 358 (1979); *Final Decision*, 77 FCC 2d 384(1980); *on reconsid*, 84 FCC 2d 50(1980); *on further reconsid* 88 FCC 2d 512(1981); *aff'd sub nom. CCIA v. FCC*, 693 F2d 198(1982)

CHAPTER 12

"Rostow Report"—Report of Presidential Task Force on Telecommunications Policy, May 1969

"Specialized Common Carriers"—Notice of Inquiry to Formulate Policy etc., 24 FCC 2d 318(1970); *First Report and Order*, 29 FCC 2d 870 (1971), *aff'd sub nom, Washington U & T Comm. v. FCC*, 513 F2d 1142(1975), *cert. denied*, 423 U.S. 836(1975)

"Domsat"—Domestic Communications by Satellite Facilities, Second Report and Order, 35 FCC 2d 844(1972); *reconsid*, 38 FCC 2d 665(1972)

CHAPTER 13

"FX/CCSA"—*MCI Comm. Corp. v. ATT*, 369 F. Supp. 1004(1973); *Bell Tel. Co. of Pa. v. FCC*, 503 F2d 1250(1974); *Bell System Tariff Offerings*, 46 FCC 2d 413(1974)

CHAPTER 14

"Execunet I"—*MCI Tel. Corp.*, 60 FCC 2d (1976), *revd sub nom MCI Tel Corp v. FCC*, 561 FCC 2d 365(1977), *cert denied*, 434 U.S. 1040(1978)

"Execunet II"—*MCI Tel. Corp. v. FCC*, 580 FCC 2d 590(1978), *cert denied* 439 US 980(1978)

CHAPTER 15

"Enfia"—*Exchange Network Facilities*, 71 FCC2d 440(1979)

"Docket 78-72, MTS/WATS Market Structure Inquiry—Report and Third Supp. Notice, 81 FCC2d 177(1980); *Second Report and Order*, 92 FCC2d 787(1982); *Third Report and Order*, 93 FCC2d 241 (1983); *aff'd NARUC v. FCC*, 737 F.2d 1095(1984), *cert denied*, 469 U.S. 1227(1985)

CHAPTER 16

MCI v. AT&T, 708 F.2d 1081(1983), *cert denied* 464 U.S. 891; *Southern Pacific Comm. Co. v. AT&T*, 556 F. Supp. 825(1982), aff'd 740 F.2d 980, *cert denied* 105 S.Ct. 1359; *Litton Systems, Inc. v. AT&T*, 700 F.2d 787(1985), *cert denied* 104 S.Ct. 984(1984)

CHAPTER 17

Hearings by Committee on Energy and Commerce of House of Representatives (except where otherwise noted)

Competition in Telecommunications Industry (H.R. 12323), Sept. 28-30, 1976, Serial No. 94-129 (94th Cong.)

Communications Act of 1978 (H.R. 13015), July 18-Sept. 23, 1978, Serial Nos. 95-194-200 (95th Cong.)

Communications Act of 1979 (H.R. 3333), April 24-June 28, 1979, Serial Nos. 96-121-128 (96th Cong.)

Telecommunications Act of 1980, Report to accompany H.R. 6121, Aug. 25, 1980, H.Rept.No. 96-1252 (96th Cong.); Adverse Rept. by Committee on the Judiciary, Oct. 8, 1980, H.Rept.No. 96-1252 (96th Cong.)

Proposed Antitrust Settlement of U.S. v. AT&T, Joint hearings with Comm. on the Judiciary, June 26, 28, 1982, Serial No. 97-116 (Serial No. 35) (97th Cong.); Senate Committee on Commerce, Science and Transportation, Jan. 25, 1982, Serial No. 97-92 (97th Cong.)

Status of Competition and Deregulation in the Telecommunications Industry, May 20, 27, 28, 1981; Serial No. 98-29 (97th Cong.)

Telecommunications in Transition, Report by Majority Staff, Nov. 31, 1981, Comm. Print 97-V (97th Cong.)

Telecommunications Competition and Deregulation Act of 1981 (S.898), Sen. Comm. on Commerce, Science and Transportation, June 2-9, 1981, Serial No. 97-1; Report to accompany S.898, July 1981, S.Report No. 97-170 (97th Cong.)

Monopolization and Competition in the Telecommunications Industry, Senate Comm. on the Judiciary, June 10-29, 1981, Serial No. J-97-20

CHAPTER 18

U.S. v. AT&T, Civil Action No. 74-1698, U.S. District Court for the District of Columbia, see transcript of hearings; *denial of motion to dismiss*, 524 F. Supp. 1336 (1981)

CHAPTER 19

"Modified Final Judgment (MFJ)"—Proposed MFJ, 47 Fed. Reg. 4166, June 28, 1982; Opinion of Judge Harold H. Greene, Aug. 11, 1982, 552 F. Supp. 131, *aff'd sub nom. Maryland v. U.S.*, 103 S.Ct. 1240 (1983)

CHAPTER 20

"MFJ Implementation"—*Plan of Reorganization*, 569 F. Supp. 1057 (1983); *Redefined Service Areas* (LATAs), 569 F. Supp. 990 (1983); *Routing of Unallocated Toll Calls*, 578 F. Supp. 668 (1983); *NXX Plan*, Civil Action No. 82-0192, Jan. 9, 1985

INDEX

About the Authors

FRED W. HENCK is Consulting Editor of *Telecommunications Reports.*
He was editor and president of that publication for 20 years and has written
numerous articles for the telecommunications press.

BERNARD STRASSBURG is of counsel to the firm of Ward & Mendelsohn
in Washington, D.C. Until his retirement on December 31, 1973, from a
31-year career in federal service, Strassburg served for ten years as Chief of
the Common Carrier Bureau of the Federal Communications Commission.
Prior to that he was associate bureau chief as well as head of the Commission's Office of Satellite Communications.

DATE DUE

Demco, Inc. 38-293